C. L. Carter, Jr. and Associates, Inc.

MANAGEMENT AND PERSONNEL CONSULTANTS
434 Bank Bldg.
811 So. Central Expwy.
Richardson, Texas 75080
(214) 234-3296

1990-91

"QUALITY ASSURANCE, QUALITY CONTROL & INSPECTION HANDBOOK"

REVISED & UPDATED FIFTH EDITION 1990-91

By

Dr. C. L.(Chuck)Carter, Jr., PE,CQE,CRE,CQA,CMC

First Edition 1966
Second Edition 1968
Third Edition 1979
Fourth Edition 1984
Fifth Edition 1990

Published & Copyright © 1966,1968,1979,1984,1990

By: C. L. Carter, Jr. & Associates, Inc.
    Management & Personnel Consultants
    P. O. Box 5001
    Richardson, Texas 75080
    (214) 234-3296

Library of Congress Cat. Card No. 78-78394

ISBN 1-879519-09-7

NOTES:

INTRODUCTION & INTENT

We have revised & updated the "QUALITY ASSURANCE, QUALITY CONTROL
& INSPECTION HANDBOOK" for use by ALL Quality, Reliability, Safety,
Inspection, Engineering, Manufacturing, Purchasing, Materials, Con-
tracts, Programs, Sales, Marketing, Accounting, M.I.S., Maintenance,
Human Resources, and all levels of Management from President & CEO
to Supervisors and all others who have an affect upon the Quality of
Products or Services in any way.

This revised personal pocket-sized handbook has been prepared to provide Guidance, Direction, Technical Assistance, Training, Motivation
& Development to those individuals who are involved with the Integrity of Products & Services provided by or for Commercial, Industrial,
Military, Nuclear, Food, Drug, Medical, Chemical, Textile, Automobile
and or any other scientific or consumer oriented customers on a world
wide basis.

This handbook is intended for use on a day to day basis for Personal
Planning, Career Growth & Development, Training & Motivation, Orientation & Testing of New Hires and those being considered for promotion,
refresher information and for guidance & direction during the normal
operation of your daily activities at work, at home or at play. There
are things in this handbook for your children & grand-children and to
help you through the day and or night.

This is a "Mini-Text Book" to be used for Self-Education, Training &
Development Programs for the Classroom or for the Personal Growth and
Development of the Individual, the Production Operator, the Inspector/
Auditor, the Purchasing Buyer, the Engineering or Test Technician, the
Vendor Q.A. Engineer, and so on. This handbook is for YOU! This hand-
book is used in Colleges & Universities to teach & train.

People are the 'Most Important Asset' in any company. Management must,
at some point, come to recognize this before it is too late to turn
their companies around by adopting the "TOTAL QUALITY MANAGEMENT" system of operation. Management must 'Change the way they Think, the way
they Act and the way they Operate'. They must 'Do It Now'. People want
to work for Quality Companies; for management that Cares about their
employees and about Quality, management that is Interested, Involved &
Concerned about Quality, Reliability, Safety, Service & Integrity of
the entire operation; management that considers each individual as a
'Very Important Person' empowered to do that which is right, to suggest
Quality Improvements & Changes to enhance the company, a process/part/
product or service, only for Recognition as a Contributor to the growth
& development & success of the company. There is alot of information in
this handbook for Executive Management and we hope they see & use it.

We believe it is necessary & required to have an 'On-going Career Training, Growth & Development Program' for every individual in the company
if the company expects to grow to and or maintain #1 Status in the Quality Marketplaces of the World. Unify for Quality! Its the Only Way!

                                                          Dr. Carter

QUALITY ASSURANCE, QUALITY CONTROL & INSPECTION HANDBOOK

TABLE OF CONTENTS

Fifth Edition Total Pages = 212

BOOKS, VIDEO TAPES, SLIDES & TRAINING MATERIALS
AVAILABLE FROM C. L. CARTER, JR. & ASSOC. INC.

* Quality Assurance Workmanship Standards & Training Manual

* The Control & Assurance of Quality, Reliability & Safety

* Quality Assurance, Quality Control & Inspection Handbook

* Reliable Soldering...VHS Video Tape 20 Minutes

* Wiring & Soldering Workmanship Standards Training Program..Slides 100

* Wiring & Soldering Training Program...Video Tape VHS (H)

* Soldering For Productivity: The Basics...Video Tape VHS (PE)

.......Others as Developed by Dr. Carter

Note: Some of the above are also available from THE AMERICAN SOCIETY
      FOR QUALITY CONTROL AS COVERED IN THEIR CATALOG.

ACKNOWLEDGEMENTS

Dr. Carter gratefully acknowledges the time, talent, effort
and dedicated work of his associates who have devoted long
hours to the preparation of this edition of "QUALITY ASSUR-
ANCE, QUALITY CONTROL & INSPECTION HANDBOOK". We also acknowledge the anonymous inputs which have been made & incorporated into this handbook. Thank You All Very Much!

A Personal Note: Since I am a Registered Prof. Engineer in
Quality & Manufacturing; Certified by ASQC as a CQE/CRE/CQA;
Certified by SME as a CMfg.E.; Certified by ICPM as a Professional Manager; Certified as a Management Consultant(CMC)
by IMC; and Certified by NPA as a Counseling Psychologist,
you may want to contact me concerning details about your
personal career development program. Please do! I will be
very pleased to communicate with you.

A PROFESSIONAL REFERENCE LIBRARY FOR INFORMATION

This is a 'Starting Point' for you to obtain more information on
and from SOCIETIES, ASSOCIATIONS, STANDARDS WRITING GROUPS, BOOKS,
ORGANIZATIONS, etc. that can help & assist you. This has been developed by Dr. Carter over a period of time and up-dated with each
new revised edition of this handbook. The addresses are accurate
to the best of our knowledge as of this revised edition.

1. C. L. Carter, Jr. & Associates, Inc.
   Quality Assurance & H. R. Consultants
   P.O.Box 835001
   Richardson, Texas 75083
   (214) 234-3296
   Quality Books & Training Materials; Consulting & Training Services;
   Career Counseling & Evaluation. See Books, Videos, Slides, below:
     Quality Assurance Workmanship Standards & Training Manual...Book
     The Control & Assurance of Quality, Reliability & Safety....Book
     Quality Assurance, Quality Control & Inspection Handbook....Book
     Reliable Soldering...VHS Video Tape for Solder Cert. Training
     Soldering For Productivity...VHS Video for Solder Cert. Training
     Wiring & Soldering Training Program...VHS Video " "   "   "
     Wiring & Soldering Workmanship Stds. Training Program...100 Slides

2. American National Standards Institute   (Catalog of Stds. on Qual.
   1430 Broadway   Safety, etc. ISO/ANSI/ASQC
   New York, N.Y. 10018   9000 & Q90 Series,PLUS..)

3. American Society for Quality Control   (Prof.Society; Cert.Programs
   310 West Wisconsin Ave.   for YOU; Magazine/Pubs.;
   Milwaukee, Wisconsin 53203-2213   Books; Quality Press;Have
   (414) 272-8575 Fax:(414) 272-1734   Divisions & Tech. Comm.)

4. European Organization for Quality Contl.   (Prof.Society; EOQC Magazine;
   P.O.Box 5032   Publications)
   CH-3001, Berne, Switzerland

5. American Society of Safety Engineers   (Prof.Society; Magazine &
   850 Busse Highway   Pubs.; Cert.Program CSP)
   Park Ridge, Illinois 60068

6. Test Magazine   (All types of testing is
   61 Monmouth Rd.   Covered)
   Oakhurst, New Jersey 07755

7. American Society for Testing & Materials   (Prof.Society; Directory
   1916 Race St.   of Testing Labs.)
   Philadelphia, Penna. 19103

8. Association for Quality & Participation   (Prof. Assoc.; Magazine &
   801-B West Eighth St. Suite 301   Pubs.; Local & National
   Cincinnati, Ohio 45203   Meetings; Training)

8.1 Hogan & Associates, Inc.   (Texas Quality Consortium;
    16479 Dallas Parkway Suite 390   Training Courses/Programs;
    Dallas, Texas 75248   Consulting Services)
    (214) 931-7597   Fax(214) 407-0796

9. Malcolm Baldrige National Quality Award   (Applications & Info.)
   United States Department of Commerce
   National Institute of Standards & Technology
   Gaithersburg, Maryland 20899
   (301) 975-2036 Fax(301) 948-3716

10. Quality Digest Magazine   (People Oriented Magazine;
    QCI International, Dept. 3000   Quality Articles by Crosby,
    P.O.Box 1503   Peters, Iacocca; Well-done
    Red Bluff, Calif. 96080   work by staff/contributors)

11. Underwriters Laboratories (U/L)   (Testing & Qualification
    Chicago, Illinois   Labs.; ISO 9000 Cert/Reg.)

12. American Bureau of Shipping (ABS)   (Technical Services;Insp.;
    ABS Quality Evaluations, Inc.   Verification;Evaluations &
    263 No. Belt East   Certifications per ISO
    Houston, Texas 77060   9000 Standards/Registration

13. National Association of Manufacturers   (Prof. Association)
    277 Park Ave.
    New York, N.Y. 10017

14. Society of Manufacturing Engineers   (Prof.Society; Magazine &
    One SME Drive, P.O.Box 930   Pubs.; Books & Videos)
    Dearborn, Mich. 48121

15. Administrative Management Society   (Prof.Society; Local &
    AMS Bldg., Maryland Road   Regional Meetings; Cert.
    Willow Grove, Penna. 19090   Admin. Mgr. Program)

16. American Society of Mechanical Engineers   (Prof.Society; Standards;
    345 East 47th St.   Specifications, etc.)
    New York, N.Y. 10017

17. Society of American Value Engineers   (Prof.Society;Cert.Prog.)
    29551 Greenfield Rd. Suite 210
    Southfield, Mich. 48076

18. American Society for Training & Devel.   (Prof. Society; Training &
    1630 Duke St., P.O.Box 1443   Devel. Journal/Magazine;
    Alexandria, Virginia 22313   Tech/Skills Train Mag)

19. Canadian Standards Association   (Canadian Quality Standards
    178 Rexdale Blvd.   Z299, etc.; Certification
    Rexdale, Ontario, Canada M9W 1R3   of Products & Companies to
                                                Z299 & ISO 9000 Standards

20. Institute of Elec/Electronic Engineers   (Prof.Society; Stds/Pubs.;
    345 East 47th St.   Software Standards)
    New York, N.Y. 10017

20.1 The Carman Group, Inc.   (Cost of Quality S.W.;
     P.O.Box 867689   Training Seminars)
     Plano, Texas 75086
     (214) 867-5089   Fax (214) 669-9478

21. National Tooling & Machining Assoc.   (Apprenticeship & Train-
    9300 Livingston Rd.   ing Programs)
    Washington, D.C. 20022

22. State Board of Registration for   (Prof. Registration of
    Professional Engineers-California   Quality, Mfg.; Safety
    1006 Fourth St.   Engineers & others)
    Sacramento, Calif. 95814
    (916) 445-5544

23. National Society of Prof. Engineers   (Prof.Society for P.E.'s;
    1420 King St.   Magazine; Pubs.; Books &
    Alexandria, Virginia 22314-2715   Videos)

24. Superintendent of Documents   (MIL-Stds & Specs; Pubs.)
    U.S. Government Printing Office
    Washington, D.C. 20402

25. National Management Association   (Prof.Assoc.; Training Pro-
    2210 Arbor Drive   grams; Cert.Prof.Mgr. via
    Dayton, Ohio 45439   Inst. of Cert. Prof.Mgrs.)

26. American Society for Non-Destructive Test.   (Prof.Society; Materials
    4153 Arlingate Plaza, Caller#28518   Evaluation Magazine)
    Columbus, Ohio 43228

27. National Association of Purchasing Mgt.   (Prof.Society; Magazine &
    P.O.Box 78059   Pubs.; Cert.Purch.Mgr. &
    Phoenix, Arizona 85062-8059   Training Programs)

28. American Management Association   (Prof.Assoc.; Pubs.; Train-
    135 West 50th St.   ing Courses/Seminars)
    New York, N.Y. 10020

29. Institute of Management Consultants   (All Members are Certified
    230 Park Ave.   Management Consultants-
    New York, New York 10169   CMC; Dr. Carter is member)

30. Computer & Automated Syst. Assoc. of SME   (Prof.Assoc.; Magazine &
    One SME Drive, P.O.Box 930   Pubs.; Conferences
    Dearborn, Mich. 48121

31. American Assoc. for Counseling & Development (Prof.Assoc.; Divisions;
    5999 Stevenson Ave.   Publications/Books)
    Alexandria, Virginia 22304

32. Texas Assoc. for Counseling & Devel.   (Prof.Assoc.; Local & Re-
    316 W. 12th St. Suite 402   gional Meetings/Training)
    Austin, Texas 78701

33. Quality Control Handbook and Others   (From McGraw-Hill; ASQC-
    By: Dr. J. Juran   Quality Press/Book-stores)

34. Out of the Crisis and other Books   (From ASQC Quality Press/
    By: Dr. W.E. Deming   Book Stores)

35. Quality is Free and other Books   (From ASQC Quality Press/
    By: Phil Crosby   Book Stores)

36. Reliability Handbook, Edited   (McGraw-Hill; ASQC Quality
    By: Grant Ireson   Press)

37. Quality Assurance: Management & Technology & Others
    By: Dr. Glenn Hayes   (From ASQC Quality Press)

38. Quality Control   (From Iowa State Univ.
    By: R. C. Vaughn   Press, Ames, Iowa 50010

39. Handbook of Systems & Product Safety   (From Prentice-Hall, Inc.
    By: Willie Hammer   Englewwod Cliffs,N.J.

40. Total Quality Control & Other Books   (From ASQC Quality Press)
    By: Dr. A.V. Feigenbaum

41. In Search of Excellence & other Books   (From Harper & Row; Book
    By: Tom Peters   Stores)

42. Process Quality Control & other Books   (From ASQC Quality Press)
    By: Dr. Ellis Ott

43. Quality Magazine   (Free Publication-Monthly)
    Hitchcock Publishing Co.
    Hitchcock Bldg.
    Wheaton, Illinois 60187

44. Dray Publications, Inc.   (Safety & Health Pubs.)
    Deerfield, Mass. 01342

45. Clemprint, Inc.   (Quality/Safety Poster
    Concord Industrial Park   Programs)
    Concordville, Penna. 19331

46. Training Magazine   (Human Resources Magazine)
    Lakewood Publications
    731 Hennepin Ave.
    Minneapolis, Minn. 55403

47. Circuits Manufacturing Magazine & Others (Printed Circuits, etc.)
    Benwell Publishing Corp.
    1050 Commonwealth Ave.
    Boston, Mass. 02215

48. The Industrial & Process Control Magazine (Industrial Process Con-
    Chilton Company   trols)
    Chilton Way
    Radnor, Penna. 19089-0380

49. Intertek Services Corp.   (Inspection, Surveys &
    655 Deep Valley Dr.   Auditing Services)
    Rolling Hills, Calif. 90274

50. Southwest Research Institute   (Q.A. Testing for Business,
    P.O.Drawer 28510   Industry & Government)
    San Antonio, Texas 78284

QUALITY PERFORMANCE -- ALWAYS GOES STRAIGHT TO THE HEART!

QUALITY, RELIABILITY AND SAFETY IS EVERYBODY'S BUSINESS!

EDUCATION: 1990 - 2010

Education in 1990 is being described as 'THE PITS'.....
and in general this may be true in the larger cities &
states with internal 'People Problems' from politicians
in office to parents in homes who appear to have lost
many or all of their personal values and their integrity.
We need to CHANGE from BAD to GOOD. We need to do it NOW!!

As we enter a decade of intense global competition in the
world of Business, Science and Technology, America's schools
are fighting an uphill battle to educate our children. We
all know the reasons: Low Academic Standards; Lack of Fund-
ing; Poorly Motivated Teachers; Outmoded Equipment; Crowded
Classrooms; Drug Abuse; Broken Homes; Gangs & Gang Wars;
Teenage Pregnancy; Greed; Misuse & Abuse of Power; I Don't
Care Itis; Drop-outs & Graduates can't read, comprehend,
add, subtract, divide or multiply and can't perform on the
job without extensive & costly 'Rework Training'by employers
to try to bring them up to minimum performance levels...and
the list goes on and on......

In a recent achievement test as given to high school stud-
ents   in 13 countries, the United States scored 11th in
chemistry, 9th in physics and last in Biology. What does
that say about our ability both now and in the future to
compete in the marketplaces of the world? Not real good.

We are already facing an estimated shortage of 700,000 new
scientists and engineers by the year 2010. The U.S. Department of Labor predicts that the need for higher skills in
most jobs will double in the next 10 years and this is also
supported by recent studies & predictions by The American
Society for Training & Development of which I have been an
active member for many years. We are in trouble and we have
problems to solve. The good news is that we can solve them!

Many companies rely on highly skilled innovative people to
create, design, manufacture, inspect, test, sell, service
and serve their customers. Our future as a nation depends
on a well-educated workforce. I think it is the pits when
we must import educated people from other countries because
we can't grow, develop, train & educate our own children!!
It is incumbent upon the business sector in the United States
to intensify its efforts to improve the Quality of Education
in America. Many companies already support our schools at all
levels on a ongoing basis. But more help is needed. Not just
money, but personal involvement.

We're counting on your kids and my grand-children to save
the environment, design a crash-proof car or perhaps pilot
a Spacecraft to Mars or Jupiter or just fly us to the moon!
We want our kids to graduate from good schools as taught by
dedicated caring teachers. We want our kids to be 'Somebody'
and at this point I think one of the best solutions is the
KOALATY KID PROGRAM which seeks to improve education through
Quality Improvement Techniques. Quality starts at Home and pro-
ceeds to all levels of education. Push this Program!!

THE KOALITY KID PROGRAM is geared to teaching kids to be QUALITY KIDS and it starts with the FAMILY and then starts again in Elementary School. It starts again with the Local School System from TOP to Bottom in Administration and to all Teachers. It then continues to Junior & Senior High Schools and the Junior/Community Colleges on into Senior Colleges & Universities. This type of Quality Program is in use and is now moving into Cities & State Government. The Defense Department is now stated to be committed to "TOTAL QUALITY MANAGEMENT" as the their way of life from now on. Companies are slowly moving in that direction also and it is my personal & professional opinion that this is the only way to Succeed, at work, at home and at play. Quality Improvement Methodology & Techniques will work in all these areas of our lives and we need to proceed as Quickly, Quietly & Cost Effectively as we possibly can.

The latest statistics tell us that....

1. Our school systems are in deep trouble...
2. There is no old fashion Discipline in the Home, at School (at any level), at Work or at Play.
3. Everybody is blaming everyone else...from Top to Bottom.
4. No one wants to be Responsible for their actions....
5. Cheating & Greed starts at home and is in Schools, Work, Local, State & National Government, Business & Industry.
6. Read the papers or listen to the news...they cover 1 - 5!
7. Kids can't pass tests to Graduate...they do not read well and their math skills are so poor, business & industry is spending millions to teach them what they need to know.
8. Business executives say they are worried about retaining a stable work force while their employees say that Management Talks a Good Game about Quality but does not perform as they say i.e. 36% of employees say they do not participate in Quality Programs because they are not included; & 21% say their companies do not have any common Quality Improvement Activities in place. Reference is made for your information & action in reading the 1990 ASQC/Gallup Survey which is not good. There are Wide Gaps between what Management says and what their Employees say and thats not good.
8. We all need to get to know each other better to establish high levels of TRUST if we are going to succeed in the 90's.
9. When a Judge or Jury sends a convicted person to prison for 5, 10, 15 or 20 years, they should not be getting out because the prisons are too full as it is now at least in the State of Texas.
10. It seems that no one wants to say "You Can't Do That"....
11. Things are very, very loose from Top to Bottom in every phase of most operations....Band-aid, Quick Fixes, But not enough time to really put the fire out for good and I am talking about most Small to Medium & Lots of Large Companies in this country.
12. It has to CHANGE and Change for the Better if you really want to SUCCEED in the 90's & Beyond. We can't rock along any longer. Its Discipline, Education, Training & Development with a Big Q for Quality in front of each....

Dr. Carter

"OUR CONCEPT OF QUALITY, RELIABILITY & SAFETY"

Our concept of Quality involves the entire management team from Sales and Marketing, through Engineering Design and the Design Review, to Manufacturing and Materials Management, through Quality Control and the Inspection Process, to Shipping and out to the Customer, with ongoing 'Audits' by the organization we call "QUALITY, RELIABILITY & SAFETY." We consider Quality, Reliability and Safety to be everybody's business! As such, every individual in the company must be completely involved. You and I recognize that this cannot take place by next Thursday! However, with professional assistance and with the right people in the right place at the right time, we can make it happen and make it successful and beneficial to the company and to its people.

Our concept is based on sound planning for long term growth and development of the company and all of its people by means of a career development program in concert with a Total Quality, Reliability, Safety, Training and Materials Management Program. Our concept requires management dedication, commitment, interest, involvement, concern and trust. Without these, no program can be effective over a sustained period of time. Our concept will provide and show you progress toward goals and objectives from startup to implementation and beyond. You will 'See and Feel Progress' as the Corporate Policies and Procedures are documented to meet the need of the company by our consultant(s) in concert with your management people. All policies and procedures are coordinated with management prior to implementation. Documented progress reports are provided to allow maximum communication, coordination and cooperation at all levels.

Our concept is based on participation, people, planning and profitability. Our concept is based on improving productivity with a management system of Prevention, Control, Assurance, Assistance and Auditing that is accepted and understood by the entire management and personnel team and 'Worked' as a system should be worked to yield results. Quality is a 'Gold Mine'. However, you must 'Work the Mine' if you are to achieve results and obtain the gold which is 'Profit with Quality'.

Our concept blends Quality, Reliability, Safety and Training into a centralized unit for flexibility and cost effectiveness. Quality Awareness, Quality Circles, Human Relations, Participative Management, Motivation, Changing Attitudes and other modern methodology are all utilized to the fullest for the good of the entire corporate structure. Product and People Safety are combined to assure that your people are safe and your products are safe. Our concept reduces the management risks in Product Liability. Our concept is People and Prevention Oriented. Our concept is more than a Quality Circle! We do not believe that Quality Circles alone will solve or resolve your Product and People problems. We believe our 'Concept on Quality and People' will bring about the desired results, faster, in a planned, systematic manner, at a lower total cost.

Dr. C. L. (Chuck)Carter Jr., P.E.

QUOTES TO QUOTE & THINK ABOUT....

...."You and I are responsible for the Quality of what we do on a daily basis, at Work, at Home & at Play." Dr. C. L. Carter, Jr.

...."Don't bother just to be better than your contemporaries or predecessors. Try to be better than yourself." WILLIAM FAULKNER

...."Solving a problem may be easier than you think. You just need a Systematic Approach." Dr. W.EDWARDS DEMING

...."To a remarkable degree, our lives are increasingly dependent on the Quality of Products & Services." Dr. JOSEPH M. JURAN

...."Success comes from having the proper aim as well a the right ammunition." PROVERB

...."The new-found passion for people, quality & service has flourished amid good times. It's far from certain that "Quality First," "Peerless Service" and "Self-managed Work Teams" will remain management's battle cries when(if) the yogurt hits the fan. I fear that panicky companies will put Quality Programs on hold, consider service excellence a frill, hesitate on work-team experiments and renege on Supplier "Partnerships". It'll be back to "Ship the Product," "Crack the Whip" and "Circle the Wagons." TOM PETERS

...."The success of the Malcolm Baldrige National Quality Award has demonstrated that government & industry, working together, can foster excellence." ROBERT MOSBACHER, Secretary of Commerce

...."If you really practice Quality Improvement at the grass-roots, it will reduce costs at the same time it produces Customer Satisfaction." FREDERICK W. SMITH, Chairman, Fed. Express

...."Ninety percent of what we call Management consists of making it difficult for people to get the job done." PETER DRUCKER

...."In the Brownie troop, the Army platoon or the Factory, there is no implementation until there is engagement & commitment." TOM PETERS

...."For a large organization to be effective, it must be simple. For it to be simple, its people must have self-confidence...... Frightened, nervous management uses thick, convoluted planning books.....filled with everything they've known since childhood. Real leaders don't need clutter." JACK WELCH, Chairman of G.E.

...."To compete & win, we must redouble our efforts, not only in the Quality of our goods & services, but in the Quality of our thinking, in the Quality of our response to customers, in the Quality of our decision-making, in the Quality of everything we do." E. S. WOOLARD, Chairman & CEO, E.I. DuPont

...."To me, total quality deployment is synonymous with being market driven, since our customers are the final arbiters of how good we are. Without satisfied customers, there is no bottom line." JOHN F. AKERS, Chairman & CEO, IBM

NATIONAL QUALITY MONTH

National Quality Month was launched in 1984 with a Joint Resolution
of Congress and a Presidential Proclamation designating **October** as
National Quality Month. The original legislation was reaffirmed in
1988 by the U.S. Congress and President Bush.

National Quality Month is organized & directed by The American Society for Quality Control. Its purpose is to: "Increase Awareness of
Quality Improvement as a Primary Means by which Business & Industry
can Improve the Quality of Products & Services, Increase Profits
and Strengthen its position in The Marketplaces of the World.

National Quality Month 1990 officially began on October 2nd with
the 6th annual National Quality Forum as held in New York City.
The 1990 National Quality Month's theme was "The Human Side of
Quality Improvement". The sold-out Forum brought together leaders
in business & industry along with academia, to discuss the latest
Quality Issues.

The Forum was Televised Closed Circuit around the world and to
specific sites in the U.S.A., Canada, etc. where local ASQC Sections and businesses welcomed members and/or employees to share
the live Forum from New York City. There were 1,720 Satellite Sites.

Yes, Quality is Catching-on....An example being in Dallas, Texas.
The 1989 Forum along with a day long Quality Program had about
500 in attendance. The 1990 Forum & Quality Program had 1000 in
attendance. Many other ASQC Sections are now celebrating National
Quality Month not just on the day of the Forum, but All Month!

Comment by Dr. Carter: **Everyday must be or become "A Quality Day"
365 Days a Year IF SMALL, MEDIUM & LARGE Business & Industry is
to survive. We are making progress, but we have a long, long way
to go just to get most firms to understand what Quality Means &
What Quality Can Do For Them for both the short & long term....
Quality Professionals must & will continue to "PREACH, TEACH,
SHOW & TELL, UNTIL WE HAVE PUT OUR ARMS AROUND THEM ALL".**....

WORLD QUALITY DAY

A DAY IN
November
has been agreed
for World Quality Day
EACH YEAR..... IN 1990 it was November 8th.

Full Member Organizations of EOQ are taking all appropriate actions to ensure that this day is recognized, publicized and treated in such a way to provide a high profile for quality in the broad sense.

The European Organization For Quality of which I am a Member, selects a specific day each year as WORLD QUALITY DAY...

Dr. C. L. Carter, Jr. P.E.

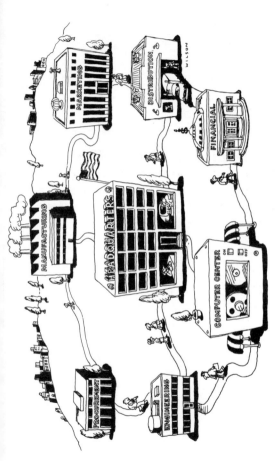

EVERYONE IS INVOLVED IN QUALITY...EVERYONE MUST COME TO UNDERSTAND THIS CONCEPT & PHILOSOPHY OF OPERATION REGARDLESS OF THE TYPE OF BUSINESS, LARGE OR SMALL, PRODUCT OR SERVICE ORIENTED.

Dr. C. L. (Chuck) Carter Jr., P.E.

A company without effective long-range planning is like a ship without a rudder. Sooner or later it drifts onto the rocks of hard times.

Characteristic Weaknesses of Troubled Companies

MANAGEMENT SHORTCOMINGS
- Critical skills not on board
- One-dimentional management emphasis
- Failure to delegate effectively
- Lack of accountability for results
- Stretched to limits of capacity

WEAK LINKS TO MARKETS & CUSTOMERS
- Product-oriented vs. customer-oriented
- Weaknesses in after-sale support
- Limited and ineffective customer feedback
- Frequently surprised by competitors' actions

LACK OF WELL DEFINED STRATEGY & FOCUS
- Narrow view of markets and business opportunity
- Lack of consistency in implementation
- Targets of opportunity vs. strategic objectives
- Short-term focus: survival vs. strategy
- Objectives and strategies not broadly communicated

ELEMENTS OF INTERNAL STRUCTURE WEAK OR MISSING
- Timely and relevant management information
- Accounting, cash management & financial controls
- Purchasing & inventory management
- Order entry & production scheduling
- Quality management & controls
- Planning/performance measurement process

FINANCIAL CONSTRAINTS & LIQUIDITY PROBLEMS
- Pattern of operating losses
- Failure to manage working capital assets
- Weak relationships with financial sources
- Inadequate capitalization

VULNERABILITY TO EXTERNAL FACTORS
- Narrow product/market base
- Limited market intelligence or customer feedback
- Dependency on single customer or supplier
- Weak financial structure
- Organizational inertia - unable to react or adapt to change

Reference: Dallas Business Journal
            Joseph Picken

Comment by Dr. Carter: This is an Excellent Analysis in
                       Every Respect and I Agree Completely.

MATURITY....THE ABILITY TO SEE THE BIG PICTURE

Maturity is many things. First, it is the ability to base
a judgement on 'The Big Picture'...the 'Long Haul'. It means
being able to pass up the fun-for-the-minute and select the
course of action that will pay off later. One of the characteristics of infancy is the 'I-want-it-now' approach. Grown-
up people are able to wait.

Maturity is the ability to stick with a project or a situation until it is finished. The person who is constantly changing jobs, changing mates and changing friends is immature.
Everything seems to turn sour after awhile.

Maturity is the ability to face unpleasantness, frustration,
discomfort & defeat without complaint or collapse. The mature
person knows he or she can't have everything his or her own
way. Nobody wins them all. He is able to defer to circumstances,
to other people....and to time.

Maturity means doing what is expected of you, and this means
being dependable. It means keeping your word. Bound in with dependability is personal integrity. Do you mean what you say...
and say what you mean? Do you do what you said you were going to
do like you said you were going to do it!

The world is full of people who can't be counted on. They are
never around in a crisis. They break promises & substitute alibis
for performance. They show up late....or not at all. They are con-
fused & disorganized. Their lives are a maze of unfinished business.
Such behavior suggests a lack of self-discipline...which is a large
part of maturity.

Maturity is the ability to make a decision and stick with it,
riding out the storms that may follow. This requires clear thinking. And the courage to stand by your position once you've taken it.

Immature people spend a lifetime exploring possibilities and then
doing nothing. Action requires courage. And courage means maturity.

Maturity is the ability to harness your abilities and your energ-
ies and do more than is expected. The mature person refuses to
settle for mediocrity. He or she would rather aim high and miss
the mark than aim low...and make it.

Author Unknown

<u>Comment by Dr. Carter</u>: We need mature people who care & are concerned
about people. Mature people motivate others. Maturity & Motivation go
hand in hand and are catching!

Are You a Winner . . . or a Loser? . . .

The people in every field of endeavor can generally be identified as WINNERS or LOSERS. Few are found in the middle.

This label was first applied to coaches of athletic teams, then to individual athletes, and so on.

People in loss prevention, whether a safety director of a corporation or an insurance representative cutting his eye teeth in field inspections, are also so categorized. Associates and bosses do not have to wait for "the final score" to ascertain whether an individual is a winner or a loser. It is projected . . . in fact broadcasted . . . in every action.

During the early sixties there were three regular repeat winners on the professional golf tour. In fact Palmer, Player and Nicklaus were known as "The Big Three." If anyone else won a major event it was considered freaky or lucky.

Billy Casper tired of entering tournaments under the illusionary handicap of, "The Big Three and Me." Determined to make it "The Big Four," he posted the following where he would read it frequently:

1. A WINNER says "Let's find out," a LOSER says "Nobody knows."
2. When a WINNER makes a mistake, he says, "I was wrong." When a LOSER makes a mistake, he says, "It wasn't MY fault."
3. A WINNER goes through a problem . . . a LOSER goes *around* it and never gets *past* it.
4. A WINNER makes *commitments* . . . a LOSER makes *promises*.
5. WINNERS think, "I'm good, but not as good as I can be." LOSERS rationalize that, "I'm not as bad as a lot of other people."
6. A WINNER tries to *learn* from those who are above him, while LOSERS try to find fault with them.
7. A WINNER says, "There ought to be a better way to do it;" a LOSER says, "That's the way it's always been done here."

As the Ten Commandments can be capsulated into one Golden Rule, so can the seven above:

Winners perform; losers talk

The above contributed by Charles Culbertson, director of loss prevention for Marriott Corp.

"Check it! Check it! There's gotta be a bug!"

AS A MANAGER.... AS A PROSPECTIVE MANAGER.....

YOU WILL BE ABLE TO CHANGE/MODIFY SOME 'ATTITUDES'....

SOME WILL BE TOUGHER THAN OTHERS.....

AS WITH SOME HORSES.... THERE ARE SOME THAT WILL NOT CHANGE,
                       SOME YOU CAN'T 'BREAK' &
                       SOME YOU WILL NEVER RIDE....

your charter is to..."STAY COOL & CALM ON TOP & PADDLE LIKE
                       HELL UNDERNEITH" TO WIN ALL YOU CAN...
                       FOR THE "TOTAL QUALITY TEAM"

'Quality Management Seminar" By: Dr. Carter

10 Keys to Effective Listening

Take the time, at the beginning, to convey the idea that listening plays a
vital role in learning. This means not only listening to you, but to each other.

THE 10 KEYS	THE BAD LISTENER	THE GOOD LISTENER
Finds areas of interest.	Tunes out dry subjects.	Is opportunistic; asks, "What's in it for me?"
Judge content, not delivery.	Tunes out if delivery is poor.	Judges content, skips over delivery errors.
Hold your fire.	Tends to enter into argument.	Doesn't judge until comprehension complete.
Listen for ideas.	Listens for facts.	Listens for central themes.
Be flexible.	Takes intensive notes, using only one system.	Takes fewer notes. Uses 4-5 different systems, depending on speaker.
Work at listening.	Shows no energy output; attention is faked.	Works hard; exhibits active body state.
Resist distractions.	Distracted easily.	Fights or avoids distractions, knows how to concentrate.
Exercise your mind.	Resists difficult expository material; seeks light, recreational material.	Uses heavier material as exercise for the mind.
Keep your mind open.	Reacts to emotional words.	Interprets color words; does not get hung up on them.
Capitalize on fact that thought is faster than speech.	Tends to daydream with slow speakers.	Challenges, anticipates, mentally summarizes, weighs the evidence, listens between the lines to tone of voice.

You must 'Learn to Listen to Learn'......Dr. Carter

HOW WE LEARN

1% through **TASTE**
1.5% through **TOUCH**
3.5% through **SMELL**
11% through **HEARING**
83% through **SIGHT**

METHODS OF COMMUNICATION	RECALL 3 HOURS LATER	RECALL 3 DAYS LATER
A. Telling when used alone	70%	10%
B. Showing when used alone	72%	20%
C. When a blend of telling and showing is used	85%	65%

"THE TRAINING & DEVELOPMENT OF YOUR PEOPLE"

Training the Production Operator/Technician and the Quality Inspector/
Auditor in a planned & systematic 'Training Course' as professionally
designed, developed & conducted for your Wiring, Soldering, Assembly &
Inspection People will yield Safe, Reliable, Quality Products, on time,
at the lowest total cost to your company. Warm bodies produce scrap,
rejects, rework, increased costs and customer complaints. I strongly
recommend "The Training & Development of your People". They are the
most important asset you have. If you will train them and work with
them, they will perform and take care of you.

In most companies a new employee is sent directly to the line to learn
what to do and how to do it. The supervisor or an untrained lead-opera-
tor is supposed to 'Train the New Employee'. In most cases, this does
not work out well at all because the new person is pressured to perform
in an unfamiliar atmosphere which causes 'Turn-over' and or frustration
or both. This situation does not produce safe, reliable, quality parts
or products. This management mode of operation is very costly.

The cost-effective course to take will see you sending your new hires
directly into your Production-Paced Training Class for Prospective Talent who will learn to Wire, Solder, Assemble & Inspect their work and
your parts & products....at the lowest total cost and in the minimum
period of time. You will have trained people going to the line and a
receptive supervisor who understands the program the person has just
completed as being one that will yield your company parts & products
that are right the first time. The people will have greater respect for
the company, the supervisor, and have confidence in their own ability
and skills as a trained & qualified individual who can perform. Only
trained & qualified people go on the production line. Only trained and
qualified people inspect & test the parts & products. The company benefits and the people benefit. You have a 'Quality Team' of people who
know how and who want to produce safe, reliable, quality products.

A practical professional training program will:

  * Increase Productivity   * Reduce Learning Time   * Reduce Turnover
  * Improve Methods   * Reduce Accidents   * Reduce Costs
  * Decrease Absenteeism   * Reduce Supervision   * Improve Quality
  * Improve Job-satisfaction   * Improve Communications * Reduce Scrap
  * Reduce Rejects/Rework   * Improve Reliability   * Improve Service
  * Improve Mfg. Yields   * Improve Morale   * Improve Profit

<u>Note by Dr. Carter</u>: As a Trainer, Teacher, Counselor, Communicator,
Manager & Developer of People at all levels, I urge you to Establish &
Implement a Training & Development Program as tailored to your needs.
Experience indicates that you will benefit along with your people and
your customers. Invest your money in your people...Train Them Today!

TRAINING IS.....

The Right Training Program Saves You Money Time After Time.....
But it Costs You Money Only Once!

Training is PREVENTION
Training is CONTROL
Training is ASSURANCE
Training is DEVELOPMENT
OF PEOPLE....YOUR MOST IMPORTANT ASSET!!

MOTIVATION

THE GOOD MANAGER MOTIVATES BECAUSE OF HIS KNOWLEDGE, SKILL,
AND ATTITUDE.

PEOPLE WILL RESPOND TO THE LEADER - TO THE MANAGER THAT HAS
A GOOD BALANCE OF KNOWLEDGE SKILL AND A POSITIVE ATTITUDE
IN THE MANAGEMENT FIELD. THEY WILL RESPECT THIS INDIVIDUAL
BECAUSE HE OR SHE IS QUALIFIED.

A PERSON CAN MOTIVATE BY POSSESSING A REASONABLE BALANCE OF
KNOWLEDGE, SKILL AND A POSITIVE ATTITUDE TO ALLOW HIM TO CON-
DUCT HIMSELF IN A MANNER THAT WILL COMMAND RESPECT AND WILL
CREATE IN HIS OR HER SUBORDINATES AND SUPERIORS, THE DESIRE
AND THE NEED TO ACTIVELY RESPOND.

TEN RULES FOR SUCCESS

1. Find Your Own Particular Talent.
2. Be Big.
3. Live With Enthusiasm.
4. Be Honest.
5. Don't Let Your Possessions Possess You.
6. Don't Worry About Your Problems.
7. Look Up To People When You Can...Down To No One.
8. Don't Cling To The Past.
9. Assume Your Full Share Of Responsibility In The World.
10. Pray Consistently and Confidently.

"ALL I REALLY NEED TO KNOW I LEARNED IN KINDERGARTEN"

By ROBERT FULGHUM

This wonderful down home book by Mr. Fulghum is really
a genuine work of inspiration for everyone of any age
and of any faith. He says he is a Philosopher and I
think he is. However, along the way he has been a working cowboy, folksinger, IBM salesman, professional art-
ist, parish minister, bartender, teacher of drawing &
painting, and father. He has a way with words to say
the least and here are some that I enjoy every time I
see them. I hope they touch your heart as they did mine.
I encourage you to obtain and read this outstanding book.

"ALL I REALLY NEED TO KNOW about how to live and what to
do and how to be I learned in kindergarten. Wisdom was
not at the top of the graduate-school mountain, but there
in the sandpile at Sunday School. These are the things
I learned:

   Share everything.
   Play fair.
   Don't hit people.
   Put things back where you found them.
   Clean up your own mess.
   Don't take things that aren't yours.
   Say you're sorry when you hurt somebody.
   Wash your hands before you eat.
   Flush.
   Warm cookies and cold milk are good for you.
   Live a balanced life--learn some and think some and
    draw and paint and sing and dance and play and work
    some every day some.
   Take a nap every afternoon.
   When you go out into the world, watch out for traffic,
    hold hands, and stick together.
   Be aware of wonder. Remember the little seed in the
    Styrofoam cup: The roots go down and the plant goes
    up and nobody really knows how or why, but we are
    all like that.
   Goldfish and hamsters and white mice and even the little
    seed in the Styrofoam cup--they all die. So do we.
   And then remember the Dick-and-Jane books and the first
    word you learned--the biggest of all--LOOK."

These are all Great Words & Thoughts to live by and I
sincerely recommend them all to every person in every
nation of the world. Thank You Mr. Fulghum!

                Dr. Carter

RECOMMENDED BY: Dr. C. L. Carter, Jr., P.E.

OCTOBER 1990

TECHNICAL&SKILLS
TRAINING

Motorola Sets the Benchmark for Training

*The new Motorola University trains for
Six Sigma: virtual perfection in products
and processes. See page 28.*

Published by
American Society for Training and Development

News & Trends ▼ New Products ▼ You Ask/Trainers Answer

Dr. Carter's thoughts concerning his long relationship with ASQC and why you should join & become involved with this 'QUALITY TEAM' of over 70,000 members & professionals.

If your Professional Growth & Development is important to you, and I hope it is, then I am inviting you to help yourself to grow & develop through all of the activities, conferences, training courses/seminars, technical committees, divisions and the opportunities that are available to you to work with & learn from the Senior Members & Fellows of the society as they teach, train, and give of themselves in all of the various levels of leadership within the 15 Divisions as Chairpersons, Vice-Chairs, Treasurers, Secretary's and or chairs of any of the committees within these divisions & technical committees. I know, because I have been involved since 1959. My membership number is 7933 and we are now well over 70,000 and going very strong toward 100,000!

I have chaired Technical Committees, helped and watched them grow into Divisions and I'm still going strong and I love every minute! In 1969 I was honored to receive my FELLOW and I have helped move others into that highly respected class of membership as I would be pleased to help you, over time, to earn your FELLOW.

For the past 2 years I have been privileged to be the Chairman of THE "INSPECTION DIVISION", one of the largest in the society. I am one of the CHARTER MEMBERS of the division. This past year The VENDOR-VENDEE TECHNICAL COMMITTEE (of which I am a Past Chairman) elevated itself to Division Status and is now THE CUSTOMER-SUPPLIER DIVISION. I am a CHARTER MEMBER and very proud to be. Now, I am but one member of this society and my intent here was to show you what has been available to me and what continues to be available to you.

However, there is more... Through my relationship with this society, I have earned Professional Certifications as CQE, CRE, and this past June, 1990 I earned my CQA. I am pleased to say that Mr. John Jennings, who has worked with me in two companies has earned all FIVE (5) Certifications and all who know him are justly proud of him. My message here is that You Can Do The Same Thing! Call or write and let me help you to Grow & Develop along with the thousands of others who will also want to help.

Here is the list of the current 15 ASQC Divisions:

* 1. Quality Management	* 9. Inspection
* 2. Aviation, Space & Defense	10. Biomedical
3. Automotive	11. Energy
4. Chemical & Process Ind.	* 12. Statistics
* 5. Electronics	* 13. Human Resources
6. Textile/Needle Trades	14. Software
* 7. Food/Drug/Cosmetic	* 15. Customer-Supplier
* 8. Reliability	* Dr. Carter is a Member

Quality is in your Future, regardless of your position! Let us Help!

"Organizations are more than technology,
they're PEOPLE..."

THE ADVANCED MANAGEMENT PROGRAM THEME

MANAGEMENT . . . . A PREVENTION, CONTROL, ASSURANCE, AUDIT PHILOSOPHY

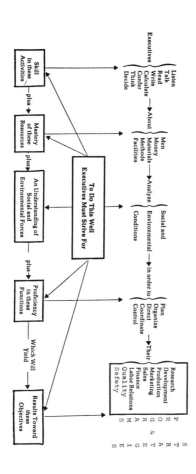

HOW QUALITY, RELIABILITY & SAFETY INTERFACES & INTERACTS AS MEMBERS OF

THE PROFIT & PRODUCTIVITY TEAM

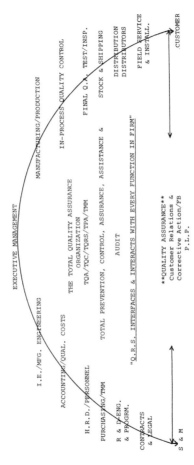

EXECUTIVE MANAGEMENT

MANUFACTURING/PRODUCTION

IN-PROCESS QUALITY CONTROL

FINAL Q.A. TEST/INSP.

STOCK & SHIPPING

DISTRIBUTION
DISTRIBUTORS

FIELD SERVICE
& INSTALL.

CUSTOMER

THE TOTAL QUALITY ASSURANCE
ORGANIZATION
TQA/TQC/TQRS/TPA/TMM

TOTAL PREVENTION, CONTROL, ASSURANCE, ASSISTANCE &

AUDIT

"Q.R.S. INTERFACES & INTERACTS WITH EVERY FUNCTION IN FIRM"

QUALITY ASSURANCE
Customer Relations &
Corrective Action/PB
P.L.P.

I.E./MFG. ENGINEERING

ACCOUNTING/QUAL. COSTS

H.R.D./PERSONNEL

PURCHASING/TMM

R & D-ENG.
& PROGRM.

CONTRACTS
& LEGAL

S & M

"ORIENTATION, INDOCTRINATION & TRAINING PROGRAM FOR MANAGEMENT PEOPLE"

Dr. Carter
2/84

<u>WHAT MANAGEMENT NEEDS TO KNOW</u>

1. **Professional** Talent is Required to do a
    Professional Job......

2. **A** Documented & Functional Prevention,
    Control, Assurance & Audit Program is
    necessary today, if you intend to compete
    in the marketplace in the future.....

3. Don't Fight the Prevention, Control, Assur-
    ance & Audit Program.....Use It!

4. You Need A Total Quality, Reliability and
    Safety Management Team for Success.......

5. **Train** & **Motivate Everyone** at **Every Level**
    Every Day..........

6. You Must Have Achievable, Attainable
    Workmanship Standards to **Consistently**
    Build Quality Products, On Time, At The
    Lowest Total Cost.....and To Calibrate
    The Eyeballs of Your People......

7. M.I.L.T.B.P.

8. Management Is Responsible For Quality,
    Reliability and Safety. Thus, You Must
    Make Each Person in Your Company Respon-
    sible For What They Do Every Day in Every
    Way......That Means Total Involvement!
    That Means:

    A) Engineering is Responsible for Q.R.S.,
       Health, Pollution, Environmental Control,
       Specifications, Design Review, to name a
       few........Quality is to Assist Engineering!
    B) Purchasing is Responsible for Q.R.S.,
       with regard to Purchasing Quality Products
       and Services on Time, at the Lowest Total
       Cost to the Company......
    C) Personnel is Responsible for Q.R.S.,
       with regard to Hiring The Right Person
       For The Right Job and Providing Assistance
       To All Departments and To All People......
    D) You Have A Professional Quality Staff.....
       Providing Assistance To Your Total Team!

Dr. Carter on....

How to
Improve
Your
PRODUCTIVITY

.......IMPROVE MARKET-SHARE & PROFITABILITY

.......IMPROVE YOUR QUALITY, RELIABILITY & SAFETY

.......PREVENT PRODUCT LIABILITY CASES & OTHER PROBLEMS

.......REDUCE PEOPLE PROBLEMS & TURNOVER

START YOUR "<u>TOTAL PREVENTION, CONTROL, ASSURANCE & AUDIT PROGRAM</u>'

......."<u>TOTAL QUALITY, RELIABILITY, SAFETY & MATERIAL MANAGEMENT</u>'

YOU WILL HAVE GREAT DIFFICULTY STAYING IN THE MARKETPLACE IN THE
1990's WITHOUT A TOTAL PEOPLE ORIENTED PROGRAM. PRODUCTIVITY WILL
COME....AFTER YOUR PEOPLE PERFORM, NOT BEFORE! YOUR PEOPLE WILL
NOT PERFORM IN TOTAL UNTIL YOU PROVIDE THEM WITH A TOTAL MANAGEME
PROGRAM THAT THEY CAN BELIEVE IN....THUS, MANAGEMENT MUST PERFOR
<u>FIRST</u> AND PROVIDE THE DYNAMIC LEADERSHIP...THE MANAGEMENT POLICY
THE MANAGEMENT PROCEDURES THAT ARE SIMPLE & CLEAR & ATTAINABLE..
THE MANAGEMENT INTEREST, INVOLVEMENT, CONCERN & DEDICATION...
MANAGEMENT MUST LEAD BY FOLLOWING THE PROGRAM, POLICIES, PROCEDU
AND IN FACT PERFORM BY DOING WHAT THEY SAID THEY WERE GOING TO D
LIKE THEY SAID THEY WERE GOING TO DO IT! UNTIL THE PROGRAM HAS T
MANAGEMENT LEADERSHIP....UNTIL THE TOTAL P.C.A.A.PROGRAM IS FULL
ADMINISTRATED AND IMPLEMENTED ON A PLANNED & SYSTEMATIC BASIS TH
THE PEOPLE CAN REALLY SEE....YOUR PROBLEMS WILL MULTIPLY DAILY!!

A TYPICAL Q.A. STANDARD for
a Quality Assurance Program
that your firm would need to
comply with if you were in the
Commercial, Industrial, Medical
field of work.

ANSI STD.Z-1.8

American National Stds. Inst.

ASQC STANDARD C1

General Requirements For
A Quality Program

Ref:"THE CONTROL & ASSURANCE
    OF QUALITY, RELIABILITY
    AND SAFETY" By Carter &
    Associates, Inc.

1. DEFINITION OF TERMS

1.1 Quality Program. The system of activities established to provide a quality of product or service that meets the needs of users.

1.2 Contractor. A term used herein to designate the individual or organization on whom this Standard is imposed.

1.3 Buyer. A term used herein to designate the individual or organization that imposes this Standard on a contractor.

NOTE

When this Standard is used internally, the "contractor" may be a single shop or production group and the "buyer" may be a management or staff group empowered to specify the use of this Standard.

1.4 Inspection. The process of measuring, examining, testing, gaging, or otherwise comparing one or more units of product with the applicable requirements.

2. SCOPE

2.1 Applicability. When this Standard is prescribed or specified by contract or agreement, it provides a specification of the general requirements to be met by the quality program of a contractor or other organization. All the requirements apply to a given contract except to the extent that they are specifically deleted, supplemented, or amended in the contract.

2.2 General Purpose. This Standard requires the establishment and maintenance of a quality program by the contractor and his subcontractors to assure compliance with the requirements of the contract. The quality program including its procedures and operations, shall be documented by the contractor and shall be subject to review by the buyer's representative.

The program shall apply to the control of quality throughout all areas of contract performance including, as appropriate, the procurement, identification, stocking, and issue of material; the entire process of manufacture; and the packaging, storing, and shipping of material.

The program shall provide that, as early as possible, discrepancies (defects and program deficiencies) shall be discovered and corrective action taken.

TYPICAL Q.A. STANDARD....continued

3. REQUIREMENTS

3.1 Quality Management.

3.1.1 General. There shall be adequate planning, forceful direction, and control in the sense of measurement and evaluation of the effectiveness of the quality program.

3.1.2 Organization. Administration of the quality program shall be vested in a responsible, authoritative element of the organization, with a clear access to management. This organization shall be staffed by technically competent personnel with freedom to make decisions without hint of pressure or bias. It shall also have sufficient authority to ensure that quality requirements are consistently maintained.

3.1.3 Procedures. Written quality control, test, and inspection procedures shall be used for all pertinent operations. These procedures shall be kept current and shall be available at all locations where they will be used.

3.2 Design Information.

3.2.1 General. Design information for a product (such as drawings, specifications, and standards) shall be maintained to ensure that items are fabricated, inspected, and tested to the latest applicable requirements. In like manner, task definitions for a service shall be maintained to ensure that the services are performed and inspected to the latest applicable requirements.

3.2.2 Change Control. All changes to design information or task definition shall be processed in a manner that will ensure accomplishment as specified, and a record of actual incorporation points (by date, batch, lot, unit, or other specific identification) shall be maintained.

3.3 Procurement.

3.3.1 General. Adequate control over procurement sources shall be maintained to ensure that services and supplies conform to specified requirements, including this specification. Purchase orders (or contracts) shall be controlled to ensure incorporation of pertinent technical and quality requirements, including authorized changes. Adequate records of inspections and tests performed on purchased material shall be maintained.

3.3.2 Source Inspection. The buyer and his authorized representatives reserve the right to inspect, at the source, any supplies furnished or services rendered under this contract. Inspection at the source shall not necessarily constitute acceptance, nor shall it relieve the seller of his responsibility to furnish acceptable product. When it is not practical or feasible to determine quality conformance of purchased items, inspection at the source is authorized.

3.3.3 Fabricated Material. All purchased material shall be evaluated to assure conformance with the requirements of applicable standards and specifications. When required, shipment of materials shall be accompanied by certified test reports that demonstrate the conformance of raw material, plating, etc., to the requirements stated in the purchase order or product specification. When submission of certified test reports is not specifically required, every shipment shall be accompanied by a certificate stating that conformance to all requirements has been ascertained, that quantitative data reports are on file, and that copies of test results will be furnished on request. The validity of certifications shall be verified periodically. Provisions will be made for withholding from use all incoming supplies pending completion of each required inspection and test or receipt of necessary test reports. The seller shall be notified whenever nonconforming materials are received, and corrective action shall be initiated when warranted.

3.3.4 Raw Materials. Raw material shall normally be tested to determine conformance to applicable specifications. Unless otherwise required by the purchase order or the product specification, certified test reports identifiable with the material may be accepted in lieu of such tests. When certifications are used as a basis for acceptance, the test results shall be compared with specification requirements. Furthermore, the validity of certifications shall be periodically verified by independent testing.

3.4 Material Control.

3.4 Material Control. Adequate methods and facilities shall be established for controlling the identification, handling, and storage of raw and fabricated material. The identification shall include indications of the inspection status of the material. These controls shall be maintained from the time of receipt of the material until delivery to the customer, in order to protect the material from damage, deterioration, loss, or substitution.

3.5 Manufacture.

3.5.1 General. Sufficient control shall be maintained over manufacturing processes to prevent excessive product defectiveness and variability, and to assure conformance of the characteristics of product, which can be verified only at the time and point of manufacture.

3.5.2 Process Control. Evaluations and controls shall be established and maintained at appropriately located points in the manufacturing process to assure continuous control of quality of parts, components, and assemblies.

3.5.3 Special Processes. Adequate methods and facilities shall be provided to assure conformance with requirements for special process specifications, such as welding, plating, anodizing, nondestructive testing, heat-treating, soldering, and testing of materials. Certifications, such as those for personnel, procedures, and equipment, shall be maintained as required.

3.6 Acceptance.

3.6.1 General. Inspection and testing of completed material shall be performed as necessary to assure that contract requirements have been met. Sufficient surveillance shall be maintained over preservation, marking, packing and shipping operations to assure compliance with requirements and to prevent damage, deterioration, loss, or substitutions.

3.6.2 Sampling Inspection. Any acceptance sampling procedures that differ from those required by the contract shall afford adequate assurance that the quality meets acceptable levels, and shall be approved by the buyer.

3.6.3 Nonconforming Material. Procedures and facilities for the handling of nonconforming material shall require prominent identification of the material and prompt removal from the work area. Unless otherwise provided in the product specification, the seller may, at his option, scrap the material or request disposition instructions from the buyer.

3.7 Measuring Instruments. Validity of measurements and tests shall be assured through the use of suitable inspection measuring and test equipment of the range, validity, and type necessary to determine conformance of articles to contract requirements. At intervals established to ensure

continued validity, measuring devices shall be verified or calibrated against certified standards that have a known, valid relationship to national standards. Tooling used as a media of inspection shall be included in this program. Furthermore, every device so verified shall bear an indication attesting to the current status and showing the date (or other basis) on which inspection or recalibration is next required.

3.8 Quality Information.

3.8.1 General. Information from control areas described in Paragraphs 3.1 through 3.7 of this specification shall be systematically utilized for the prevention, detection, and correction of deficiencies in the program that affect quality.

3.8.2 Quality Control Records. For all inspections and tests, records that include data on both conforming and nonconforming product shall be maintained. A continuing review of these records shall be made, and summary information shall be reported periodically to responsible management.

3.8.3 Corrective Action. Prompt action shall be taken to correct conditions that cause defective materials. Use shall be made of feedback data generated by the customer as well as data generated internally.

4. QUALITY PROGRAM AUDITS

Quality programs will be audited by the buyer for conformance to the intent of this specification. Disapproval of the program or major portions thereof may be cause for withholding acceptance of product.

DEFINITIONS

QUALITY CONTROL: "A management function to control the quality of articles to conform to quality standards".

QUALITY ASSURANCE: "Management discipline consisting of a planned and systematic program covering ALL functions and actions necessary to provide adequate confidence that the end item or service will perform satisfactorily in actual operation, thereby assuring customer satisfaction."

RELIABILITY: "The probability that a system or part will perform a required function under specified conditions without failure, for a specified period of time."

VALUE ANALYSIS: "An organized effort directed at analyzing the function of systems, equipment, facilities and supplies for the purpose of achieving the required function at the lowest over-all cost, consistent with requirements for performance, quality, reliability, maintainability, delivery and service."

TRAINING: "A continuous program of learning to gain knowledge and improve one's ability in any subject, art or profession. Developing skills and attitudes through self-teaching, working on-the-job, workshop seminars, or formal class room instruction."

INTEGRITY: "The completeness of an entirely integrated and honest management concept of operation which includes the Policies, Plans, Personnel, Procedures, the Product and/or Service."

MAINTAINABILITY: "The capability that exists within equipment or systems through premeditated engineering effort which permits this equipment to be restored to normal operating condition/placed back in service when a failure and/or malfunction occurs. Engineering effort which allows systems, equipment and parts to be maintainable."

ADDITIONAL WORDS, TERMS, PHRASES & DEFINITIONS

SAFETY....SAFE....SAFETY PROGRAM: "Being safe from undergoing or causing hurt, injury or loss; a device on a piece of equipment to reduce hazards; to protect against failure, breakage, or accident; A Management System of Assurance to Prevent Safety Problems in the plant, office, or product, in the best interest of the worker, the customer, the company and the community."

CALIBRATION....A Calibration System or Program: "Comparison of a measurement standard or instrument of known accuracy with another standard or instrument to detect, correlate, report, or eliminate by adjustment, any variation in the accuracy of the item being compared; A System per MIL-C-45662 or ANSI Standard."

QUALITY PROGRAM....GENERAL REQUIREMENTS: "American National Standards Institue(ANSI) Standard Z-1.8, also known as ASQC Standard C-1, provides General Requirements for A Quality Program for the commercial, industrial, and service industries. Reference: "The Control & Assurance of Quality, Reliability & Safety" by Carter."

SOFTWARE QUALITY ASSURANCE....PROGRAM REQUIREMENTS: "New Q.A. Professional Field covered by requirements in MIL-S-52779 for the development, implementation and maintenance of a Quality Assurance Program for Computer Software. Q.A. is now involved in Programming and Software. Reference:"The Control & Assurance of Quality, Reliability & Safety" by Dr. Carter."

GOOD MANUFACTURING PRACTICES....GMP's by Food & Drug Administration: "By Federal Law, each manufacturer of Medical Devices is Required to develop procedures to implement FDA's Good Manufacturing Practices Regulations. These procedures will be evaluated by the FDA to determine whether a manufacturer is complying with intent of law and regulations. Also covers imported products. Reference: "The Control & Assurance of Quality, Reliability & Safety" by Dr. Carter."

CONSUMER PRODUCT SAFETY ACT....Law 92-573 dated 10/27/72: "To protect Consumers against Unreasonable Risk of Injury from Hazardous Products. Imposes heavy fines and penalties up to $50,000 to individuals, and up to $500,000 to firms; consumer can sue the firm; can be very costly to firms without Quality & Safety Programs. Law covers thousands of products. Reference:"The Control & Assurance of Quality, Reliability & Safety" by Dr. Carter."

OCCUPATIONAL SAFETY & HEALTH ACT....Law 91-596 dated 12/29/70: "A very powerful 'People Oriented Law' which covers each and every company in the United States; requires companies to do whatever they do, wherever they do it, in a Safe & Healthful Manner; carries fines to $10,000 per violation, which can get expensive."

ADDITIONAL WORDS, TERMS, PHRASES & DEFINITIONS

INTERNATIONAL STANDARDS ORGANIZATION(ISO): The 9000 Series of International Quality Management & Quality Assurance Standards as developed & produced by Technical Committee No.176(TC176) consisting of Representatives from International Countries all seeking to clearly communicate & achieve Quality internally and between countries.

AMERICAN NATIONAL STANDARD: "A Standard which has been prepared under the prescribed 'Consensus Procedures' by the American National Standards Institute(ANSI) and approved by the ANSI Board of Review. ASQC and ASTM and many other societies & associations are Approved Standards Writing Organizations. The American Society for Testing & Materials prepares Standards for Tests & Materials. Reference is made to the latest ANSI/ASQC Q-90 Series(Q-91,92,93,94) of 1987 on Quality Syst."

FEDERAL AVIATION ADMINISTRATION(FAA): THE FAA developes & requires companies to meet 15 Specific Quality Control System Requirements in addition to all of the other Quality Specifications. Reference: "The Control & Assurance of Quality, Reliability & Safety By Dr. Carter"

NUCLEAR REGULATORY COMMISSION(NRC): Quality Assurance Requirements by ANSI N-45.2 and 10CFR, Part 50, Appendix B covers Q.A. Program Requirements for Nuclear Power Plants and the 18 Requirements to be met by any of the companies involved in Nuclear Work. Very heavy & detailed. Reference: "The Control & Assurance of Quality, Reliability & Safety"

STANDARD: "A Standard serves to provide optimum overall economy, taking into account the Functional Conditions, Quality & Safety Requirements. A Standard is made up of terms and or pictures that convey the same or like meaning each time they are used by two or more people, parties, groups, companies or organizations."

PRODUCT LIABILITY: "You are responsible and liable for what you Design, Produce, Market, Sell, Distribute, Service, Maintain, etc.; involves defective or suspect products which includes hardware & non-hardware(labeling/instructions); may involve expensive 'Recall of Products' by the FDA, CPSC, Dept. of Transportation, etc."

JOB ENRICHMENT: "Humanizing the Job; the right person for the right job; the combination of interest, training, understanding and achievement; making the job and the work place more enjoyable & safe by involving the person in their world of work."

JUST-IN-TIME: "PRODUCING Quality Products & Services On Time, At the Lowest Total Cost and Delivering whatever it is either internally to the next department in the process or the finished service/product to the customer(s) Just-In-Time as they have requested."

DOCK-TO-STOCK: "Delivering Just-In-Time Products, Parts or Materials from Your Dock to Their Stock without going thru Receiving Inspection as determined by your continued ability to meet your customer's Quality Requirements and On-Time Delivery Requirements which Qualifies you as a Certified Supplier."

SPC/SQC:"Statistical Process Control & Statistical Quality Control; maintaining all of your Processes in Control with visual charts as done manually or via computer by using Statistical Techniques."

TYPICAL FUNCTIONS OF INSPECTION, QUALITY CONTROL
AND QUALITY ASSURANCE

INSPECTION – – – – – – – – – ACCEPTANCE FUNCTION

                                      Receiving Inspection

                                      In-Process Inspection

                                      Final & Finished Goods Inspection

                                      Discrepant Material Control

                                      Day-to-Day Trouble Shooting

QUALITY CONTROL – – – – – PREVENTION FUNCTION

                                      Economic Studies & Experiments

                                      Process Capability

                                      Calibration, Tools, Gages & Equipment

                                      Design of Sampling Plans

                                      Statistical Methods

QUALITY ASSURANCE – – – – – ASSURANCE FUNCTION

                                      Quality Policy, System & Procedures

                                      Quality Standards

                                      Quality Audit, System & Product

                                      Vendor Quality Assurance

                                      Market Quality Determination

                                      Customer Complaints, Failure Analysis

                                      Analysis of Data

                                      Executive Reports on Quality

                                      Training & Motivation

TYPICAL QUALITY FUNCTIONS

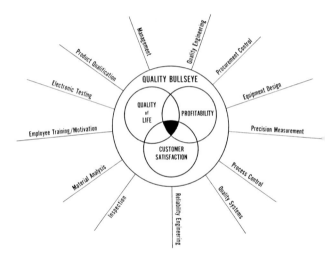

Quality Management Seminar
By Dr. Carter

"QUALITY ASSURANCE FUNCTIONS, JOBS & RESPONSIBILITIES"

1. Quality Management (Manager/Director/V.P.)

2. Quality Engineering/Software Quality Assurance

3. Procurement Control/Vendor Quality Assurance

4. Equipment Design & Test Equipment Design/A.T.E.

5. Precision Measurement & Calibration

6. Process Control/Every Process - Every Phase

7. Quality Systems/Programs/Policies/Procedures

8. Reliability Engineering/Design Review/Prevent

9. Inspection & Auditing/Any-where/Any-time

10. Material Analysis/Testing/Validation

11. Employee Training, Qualification, Motivation

12. Electronic Testing/100% Computer Controlled

13. Product Qualification/Every Part By Eng. & Q.A.

14. Configuration Control/Traceability/Recall

15. Customer Satisfaction & Corrective Action

16. Quality Cost Program/Scrap/Rework/Prevention

17. Administration, Interface & Interaction with
    All Management Functions in the Company. The
    'TOTAL PREVENTION, CONTROL, ASSURANCE, ASSIST,
    AUDIT FUNCTION'...The Total Quality Team!
    Quality, Reliability & Safety MUST BE MEMBERS
    of 'The Executive Management Team'.

By: Dr. Carter

"THE PREVENTION, CONTROL, ASSURANCE & AUDIT DOCUMENTS"

HOW CAN WE COMPLY, CONFORM & SUCCEED IN THE 1990's & BEYOND

QUALITY STANDARDS FOR CONTROLLING & ASSURING QUALITY
(Typical)

NUMBER	TITLE
1. ANSI/ASQC A1	Definitions, Symbols, Formulas, Tables Cntl.Chart
2. ANSI/ASQC A2	Terms, Symbols & Definitions for Accept. Sampling
3. ANSI/ASQC A3	Quality Systems Terminology
4. ANSI/ASQC B1.1	Guide for Quality Control Charts
5. ANSI/ASQC B1.2	Control Chart Method of Analyzing Data
6. ANSI/ASQC B1.3	Control Chart Method of Cntl. Qual. during Prod.
7. ANSI/ASQC C1(ANSI Z1.8)	Specification of Gen. Reqmts. for Qual. Program
8. ANSI/ASQC E2	Guide to Inspection Planning
9. ANSI/ASQC Q1	Generic Guidelines for Auditing Quality Systems
10. ANSI/ASQC Z1.4(105E)	Sampling Procedures & Tables for Insp. Attributes
11. ANSI/ASQC Z1.15	Generic Guidelines for Quality Systems
12. ANSI/IEEE 730	Software Quality Assurance Plans
13. ANSI/IEEE 828	Std. for S.W. Configuration Mgt. Plans
14. ANSI/IEEE 829	Std. for S.W. Test Documentation
15. ANSI/IEEE 830	Guide to S.W. Reqmts. Specifications
16. ANSI/IEEE 983	Guide for S.W. Q.A. Planning
17. MIL-Q-9858A	Quality Program Requirements
18. MIL-I-45208A	Inspection System Requirements
19. MIL-STD-45662A	Calibration System Requirements
20. Mil-STD-105E	Sampling Procedures & Tables for Insp. Attributes
21. MIL-S-52779	Software Q.A. Program Requirements
22. ANSI STD. N-45.2	Nuclear Quality Programs & 10 CFR 50 Appendix B
23. RDT F2-4T	Nuclear Inspection Systems
24. ISO/8402	Quality Assurance Vocabulary
25. ISO/9000-ANSI/ASQC Q90	Quality Management & Q.A. Standards
26. ISO/9001-ANSI/ASQC Q91	Quality Systems:Model for Q.A. in Design/Develop, Production, Installation & Servicing
27. ISO/9002-ANSI/ASQC Q92	Quality Systems:Model for Q.A. in Prod. & Install
28. ISO/9003-ANSI/ASQC Q93	Quality Systems:Model for Q.A. in Final Insp./Tes
29. ISO/9004-ANSI/ASQC Q94	Quality Mgt. & Qual. Syst. Elements - Guidelines
30. Food & Drug Laws as Appropriate to your Products	
31. Good Manufacturing Practices	
32. Guidelines for Manufacturing Safe Consumer Products: CPSCommision	
33. Canadian Standards Association/Canada Quality Council Revising CSA-Z-299-85 Series	Quality Systems to Include ISO 9000 Series(Est.'9
34. DND 1015-1016/Canadian	Same as MIL-Q-9858A & MIL-I-45662A Military
35. Consumer Product Safety Act 92-573 of 1972	
36. Occupational Safety & Health Act 91-596 of 1970: Federal & State	
37. Environmental/Pollution Laws for Air & Water & Waste: Fed/State/Local	
38. ASME Section III of Boiler & Pressure Vessel Code	
39. ASME-SPPE-1 Quality Assurance Requirements	
40. Accreditation Manual for Hospitals: Q.A. Requirements	

Note: THE MALCOLM BALDRIGE NATIONAL QUALITY IMPROVEMENT ACT OF 1987-
PUBLIC LAW 100-107 WHICH COVERS **"THE MALCOLM BALDRIGE QUALITY
AWARD"** IS FOUND IN ANOTHER AREA WITHIN THIS BOOK.

"THE PREVENTION, CONTROL, ASSURANCE & AUDIT DOCUMENTS".....Continued

    Note: Add Others as Appropriate to your Business, Products, Services, etc.
          There will be new additions to ISO/ANSI/CSA Stds. in 1991-92-93 etc.
41. MIL-DOD-STD-2167A   Defense System Software Development & H.B. 287
42. MIL-DOD-STD-2168   Defense System Quality Program & H.B. 286
43.
44.
45.
46.
47.
48.
49.
50.
51.
52.
53.
54.
55.
56.
57.
58.
59.
60.
61.
62.
63.
64.
65.
66.
67.
68.
69.
70.

    Note: All Documents Should be of the Latest Issue or Per Your Contracts

    Note: Military Standards can be obtained from THE GOVERNMENT PRINTING
          OFFICE, WASHINGTON, D.C.(There may be a Charge for the documents)
    Note: All ANSI/ASQC & ANSI/IEEE including ISO Documents can be obtained
          from THE AMERICAN NATIONAL STANDARDS INSTITUTE, 1430 Broadway.
          New York, New York 10018 (There Is a Charge for the Documents)
    Note: All ANSI/ASQC Documents can be obtained from THE AMERICAN SOCIETY
          FOR QUALITY CONTROL, 310 West Wisconsin Ave., Milwaukee, Wisconsin
          53203. (There Is a Charge for the Documents)

IN ORDER TO COMPLY WITH ANY OF THE ABOVE....(A) Obtain The Documents, NOW!
    (B) Do a Professional Review of What you Have, What you Need, etc.
    (C) Develop & Implement the System, Program, Policies, Procedures, etc.
    (D) Assure via Audits that your Total Qual. Mgt. System is Functioning
        & Providing Quality, Reliability, & Customer Satisfaction EVERYDAY!
    (E) Move to apply for Certification/Registration and when ready apply
        for 'The Baldrige Award'. With Total Commitment you can, and WIN!

                                        Dr. C. L. Carter, Jr.

TYPICAL GUIDANCE AND EVALUATION DOCUMENTS

Handbook   H50   "Evaluation of a Contractor's Quality Program"

This document provides guidance to government
personnel responsible for evaluating a contractor's Quality Program as required per
Mil-Q-9858A. It basically covers each paragraph of 9858A.

Handbook   H51   "Evaluation of a Contractor's Inspection System"

This document is used for guidance by all
government personnel when evaluating an
Inspection System as required by Mil-I-45208A.
It basically covers each paragraph of 45208A.

Handbook   H52   "Evaluation of Contractor's Calibration System"

This document provides concepts and practices
to be used in evaluating contractor's Calibration Systems as required by Mil-C-45662A.
It is for use by government personnel and
covers each paragraph of 45662A.

Handbook   H53   "Guide for Sampling Inspection"

A useful guide for quality managers, inspectors, engineers and others concerned with
basic principles of Inspection Sampling.
Amplification of Mil-STD-105 is provided.

ESSENTIAL STEPS TO QUALITY
By: Dr. Bill Wortham, P.E.

We have seen books written on Quality......
We have heard about Quality Management...... and
We have heard about "World Class Quality"......
The question still remains: "How do we improve both
our Quality and our Quality Image? For this, I don't
believe it takes a book or even a page. The answer
is as simple as 1, 2, 3...

1. Use only raw materials which meet your Standards
   and Specifications.

2. Control your Production Processes.

3. Ship only products which meet or exceed Customer
   Requirements.

Now, it may just happen that to do these things, you
will need to adopt New Principles, New Technology and
New Methods. If so, then Steps 4 & 5 must be added:

4. Change

5. Do it Now!

Special comment by Dr. Carter: Bill Wortham was a 'Special
Friend' and I had great respect and admiration for him as
the 'True Professional' he was. We consulted with one-another
and worked together as Professional Consultants. He passed-on
in July, 1990 and I will miss his counsel and his smile.

His answers are True. Do it Now!

Dr. Chuck Carter, P.E.
12/90

CHANGES ARE REQUIRED FOR PROGRESS, GROWTH, DEVELOPMENT & SUCCESS

Trained, Qualified, Skilled, Knowledgeable People are the 'Greatest Asset' to any company that plans to be in business in the 90's and beyond.... Management at all levels must be unified and make whatever changes are necessary & required to absolutely move the company to a TOTAL QUALITY MANAGEMENT SYSTEM OF OPERATION FROM TOP TO BOTTOM WITH A SIMPLE THEME OF **"Quality, On Time At The Lowest Total Cost With Complete Customer Satisfaction"**. However, Change Is Not Simple....Change Is Very Difficult To Achieve....Change Will Take Place, But Only In A Planned, Systematic Manner....Management Must Develop The Plan, Include All Employees In The Plan, Train All Employees Starting With **THE CHAIRMAN & CEO**, Recognizing That He/She or They ARE EMPLOYEES....Who Must 'Buy In & Change' Before Anyone Else Can Be Expected To Change....When Change Is Visible At All Executive Levels, Change Will Start To Take Place Through-out The Company....Change Will Not Happen Unless This Order of Change is Followed....You Will Be Spinning Your Wheels & Spending Your Money On A Lost Cause Unless You Follow The Plan & Follow The Order.... Success Can & Will Be Assured In 3 to 5, 4 to 8, 5 to 10 Years or More, As Based On Where You Are, What You Start With And How Dedicated You Are To Making IT All Happen, **"QUICKLY, QUIETLY, & COST EFFECTIVELY"**. Evaluate & Consider using 'The Baldrige Award Criteria'.

I suggest the following....

1. Adopt One of the **ISO/ANSI/ASQC QUALITY STANDARDS 9001, 9002, 9003, or Q91, Q92, Q93** as The Base-line Quality Standard for Developing & Implementing Your Management System for the Control and Assurance of Quality. There are no better Quality Standards.

2. Set your Goal to be **A QUALIFIED, CERTIFIED, REGISTERED COMPANY using a Time-Phased Plan to make it happen over a period of time (3 to 5, 4 to 8, etc.)**. This will allow you to work your way into **THE EUROPEAN MARKETS** if this is in your long range plans. If not, I can assure you that you will be required to become a QUALIFIED, CERTIFIED, REGISTERED QUALITY COMPANY to do business with your customers in the USA & Canada. **THIS IS GOING TO TAKE PLACE and I can only URGE YOU to move in this direction QUICKLY, QUIETLY & COST EFFECTIVELY**. Evaluate per Baldrige Criteria.

3. Set another Goal to put your company on a fixed course towards **A TOTAL QUALITY MANAGEMENT SYSTEM OF OPERATION** to include every Department & Function within your company. Quality must be #1 with everyone. Commit to a **TOTAL CULTURE CHANGE. Empower your People....Train your People**, all of them, in all areas of the QUALITY IMPROVEMENT PROCESS....Trust your People, communicate freely with all of your people and tell it like it is if you want them to trust you....Take care of your People & they will take care of you while saving & making more money for the company, QUICKLY, QUIETLY & COST EFFECTIVELY. Get serious about Baldrige!

4. CHANGE. DO IT NOW, SO YOU CAN PROGRESS, GROW, DEVELOP & SUCCEED. You've come a long way....Are you ready for 'The Baldrige'?
                                        Dr. C. L. Carter, Jr., P.E.

THE MANAGEMENT COMMITMENT TO TOTAL QUALITY MANAGEMENT

"MISSION STATEMENT"
(Typical)

We are in business to serve specific segments of the marketplaces of the world at a profit, by producing Quality Products and Services on time, at the lowest total cost to customers whom we know and respect and who respect us for our professionalism, competence and integrity.

We seek to achieve excellence in all facits & phases of our business. We must perform and conform to our own policies, procedures and standards which will allow us to substantially exceed the performance of our competitors, assuring the best value, quality, reliability, safety & service for our customers.

We observe high ethical standards, conduct our affairs with honesty, fairness, dignity, integrity and compassion in all relationships with our people, our suppliers and our customers. Building organizational competence is vital, and we therefore look for people whose job performance supports our corporate goal of excellence in every thing we do.

These constitute our mission to achieve excellence. Our mission is based and built upon this solid commitment to assure success.

"QUALITY POLICY"
(Typical)

WE WILL UNDERSTAND OUR CUSTOMERS AND THEIR NEEDS, MEET OR EXCEED THEIR REQUIREMENTS AND DELIVER SAFE, RELIABLE, QUALITY PRODUCTS AND SERVICES ON TIME, AT HIGHLY COMPETITIVE PRICES, TO BE 'THE STANDARD' IN THE MARKETPLACES WE SERVE, INTERNATIONALLY.

Dr. Carter
11/90

ISO 9003: QUALITY ASSURANCE POLICIES & PROCEDURES MANUAL
(Typical Contents)

Documenting the Management System for The Control & Assurance of
Quality is extremely important since Your Customers will Review
Your Policies & Procedures and then Completely Evaluate the In-
tegrity of your company with regard to your ability to conform
to your own Policies & Procedures. The same approach will be taken
by **"EXTERNAL THIRD PARTY AUDITORS AS YOU MOVE TO BECOME CERTIFIED
AND REGISTERED UNDER THE NEW REQUIREMENTS FOR CONDUCTING BUSINESS
IN EUROPE & THEREAFTER, TO CONDUCT BUSINESS IN THE U.S.A./CANADA"**.

You must now face the music and dance to the tune called 'QUALITY'.
Your company must be a Quality Company in Total. Your product(s)
must be Quality, 100% of Them, every day in every way. You must
**"Do What You Said You Were Going To Do Like You Said You Were Going
To Do It"....Without Exception!** As such, don't put things that
SOUND GOOD in your Quality Manual, if you can't perform & conform.

Here then are 'THE GENERAL CONTENTS OF A QUALITY MANUAL' where you
are a Small Business Working to Customer Drawings or Producing one
or more Standard Products Requiring Final Inspection & Test...ONLY!

1. The Responsibility of Management completely defined; clearly
   as to Who Does What, When, Where, Why & How. A Quality Policy
   and Mission Statement. Documented Reviews & Audits, etc.
2. The Management System for Inspection & Test from Receipt of
   Order, Purchasing of Materials, Parts, etc., thru all operations, processes, inspection, test, pack & ship, must be
   completely documented, flow charted, etc.
3. All procedures, drawings and related documents must be completely controlled, approved, as covered by procedures.
4. When required, the parts or products you make must be identified i.e. Batches, Lots, Part Numbers, Serialized, etc.
5. Documented procedures for Inspection & Testing must be followed,
   signed or stamped as to who did what, dated, etc. to give you
   traceability and to make the entire process auditable.
6. You must have a Documented & Functioning Calibration Program
   with all calibrations & records traceable to National/Inter-
   national Standards i.e. N.I.S.T. To give you flexibility of
   useage, I recommend that ALL INSP. & TEST EQUIPMENT i.e. Eng.,
   Production, Test & Quality, Be Calibrated at all times...
7. You must document the Status of Product concerning Insp. & Test
   by means of Stamps, Labels, Process Flow/Status Records, etc.
8. You must control all Nonconforming Parts, Materials, Products.
   A 'Management/Material Review Board' will be your best effort
   with complete documentation of Decisions i.e. Repair, Rework,
   Use-as-is; Scrap, all per documented procedures. This is for
   military & non-military companies. You need the DATA!
9. You must protect the product completely after Insp. & Test
   from any damage caused by Handling, Storage, Packaging, Delivery.
10. You must maintain Insp., Test, etc., Records from start to ship-
    ping on all products, as per contract time requirements for re-
    call on request. Note: 3,5,7 years are normal periods.
11. Your People must be Experienced or you must Train Them, Certify
    them & document everything. Train, Qualify & Certify. ItPays.
12. Statistical Quality/Process Control is Required Today. Do it!

Note: When done professionally, this would meet ISO-9003 & Mil-I-45208A

                                        Dr. C. L. Carter, Jr., P.E.

Annex

Cross-reference list of quality system elements

(This annex is given for information purposes and does not form an integral part of the standard.)

Clause (or sub-clause) No. in ISO 9004	Title	Corresponding clause (or sub-clause) Nos. in		
		ISO 9001	ISO 9002	ISO 9003
4	Management responsibility	4.1 ●	4.1 ◐	4.1 ○
5	Quality system principles	4.2 ●	4.2 ●	4.2 ◐
5.4	Auditing the quality system (internal)	4.17 ●	4.16 ◐	–
6	Economics — Quality-related cost considerations	–	–	–
7	Quality in marketing (Contract review)	4.3 ●	4.3 ●	–
8	Quality in specification and design (Design control)	4.4 ●	–	–
9	Quality in procurement (Purchasing)	4.6 ●	4.5 ●	–
10	Quality in production (Process control)	4.9 ●	4.8 ●	–
11	Control of production	4.9 ●	4.8 ●	–
11.2	Material control and traceability (Product identification and traceability)	4.8 ●	4.7 ●	4.4 ◐
11.7	Control of verification status (Inspection and test status)	4.12 ●	4.11 ●	4.7 ◐
12	Product verification (Inspection and testing)	4.10 ●	4.9 ●	4.5 ◐
13	Control of measuring and test equipment (Inspection, measuring and test equipment)	4.11 ●	4.10 ●	4.6 ◐
14	Nonconformity (Control of nonconforming product)	4.13 ●	4.12 ●	4.8 ◐
15	Corrective action	4.14 ●	4.13 ●	–
16	Handling and post-production functions (Handling, storage packaging and delivery)	4.15 ●	4.14 ●	4.9 ◐
16.2	After-sales servicing	4.19 ●	–	–
17	Quality documentation and records (Document control)	4.5 ●	4.4 ●	4.3 ◐
17.3	Quality records	4.16 ●	4.15 ●	4.10 ◐
18	Personnel (Training)	4.18 ●	4.17 ◐	4.11 ○
19	Product safety and liability	–	–	–
20	Use of statistical methods (Statistical techniques)	4.20 ●	4.18 ●	4.12 ◐
–	Purchaser supplied product	4.7 ●	4.6 ●	–

Key

● Full requirement
◐ Less stringent than ISO 9001
○ Less stringent than ISO 9002
– Element not present

NOTES

1 The clause (or sub-clause) titles quoted in the table above have been taken from ISO 9004; the titles given in parentheses have been taken from the corresponding clauses and sub-clauses in ISO 9001, ISO 9002 and ISO 9003.

2 Attention is drawn to the fact that the quality system element requirements in ISO 9001, ISO 9002 and ISO 9003 are in many cases, but not in every case, identical.

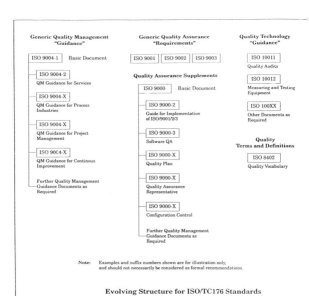

Generic Quality Management "Guidance"

| ISO 9004-1 | Basic Document |

ISO 9004-2
QM Guidance for Services

ISO 9004-X
QM Guidance for Process Industries

ISO 9004-X
QM Guidance for Project Management

ISO 90C4-X
QM Guidance for Continous Improvement

Further Quality Management Guidance Documents as Required

Generic Quality Assurance "Requirements"

| ISO 9001 | ISO 9002 | ISO 9003 |

Quality Assurance Supplements

| ISO 9000 | Basic Document |

ISO 9000-2
Guide for Implementation of ISO/9001/2/3

ISO 9000-3
Software QA

ISO 9000-X
Quality Plan

ISO 9000-X
Quality Assurance Representative

ISO 9000-X
Configuration Control

Further Quality Management Guidance Documents as Required

Quality Technology "Guidance"

ISO 10011
Quality Audits

ISO 10012
Measuring and Testing Equipment

ISO 100XX
Other Documents as Required

Quality Terms and Definitions

ISO 8402
Quality Vocabulary

Note: Examples and suffix numbers shown are for illustration only, and should not necessarily be considered as formal recommendations.

Evolving Structure for ISO/TC176 Standards

International Quality Management & Quality Assurance Standards are Produced by Technical Committee No. 176(TC176) of the International Standards Organization(ISO). These Standards are commonly referred to as the "ISO 9000 Series of Standards."

Reference: Canada Quality Council

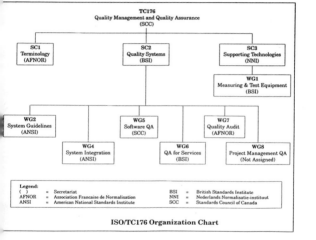

ISO/TC176 Organization Chart

International Standards Organization(ISO) & International
Electrotechnical Commission(IEC) Guides To Quality System
Registration, Certification & Accreditation

Note: You may order copies from ANSI, 1430 Broadway, N.Y.,N.Y. 10018

1. ISO/IEC Guide 2: General Terms & Definitions on Standardization.

2. ISO/IEC Guide 7: Requirements for Standards for Product Certificatio

3. ISO/IEC Guide 16: Code of Principles on Third Party Cert.Systems/Std

4. ISO/IEC Guide 22: Information on Mfgrs. Declaration of Conformity
   with Standards or other Technical Specifications.

5. ISO/IEC Guide 23: Methods of Indicating Conformity with Stds. for 3r
   Party Certification Systems.

6. ISO/IEC Guide 25: Gen. Reqmts. for Technical Competence of Test Labs

7. ISO/IEC Guide 27: Guidelines for Corrective Action to be taken by a
   Certification body in the event of misuse of its mark of conformity.

8. ISO/IEC Guide 28: General Rules for Model 3rd Party Cert. Syst./Prod

9. ISO/IEC Guide 38: Gen. Reqmts. for Acceptance of Test Labs.

10. ISO/IEC Guide 39: Gen. Reqmts. for Acceptance of Inspection Bodies.

11. ISO/IEC Guide 40: Gen. Reqmts. for Acceptance of Certification Bodie

12. ISO/IEC Guide 42: Guidelines for Step-By-Step Approach to Internatio
    Certification System.

13. ISO/IEC Guide 43: Development & Operation of Lab. Proficiency Testin

14. ISO/IEC Guide 44: Gen. Rules for ISO/IEC International 3rd Party
    Certification Schemes for Products.

15. ISO/IEC Guide 45: Guidelines for the Presentation of Test Results.

16. ISO/IEC Guide 48: Guidelines for 3rd Party Assessment & Registratio
    of a Supplier's Quality System.

17. ISO/IEC Guide 49: Guidelines for Development of Quality Manual for
    a Testing Laboratory.

18. ISO/IEC Guide 53: An Approach to the Utilization of a Supplier's
    Quality System in a 3rd Party Product Certification.

19. ISO/IEC Guide 54: Testing Lab. Accreditation Systems - General Reco
    mendations for the Acceptance of Accreditation Bodies.

20. ISO/IEC Guide 55: Testing Lab. Accreditation Systems - General Reco
    mendations for Operation.

Note: All Guides to be of the Latest Issue by Date.

QUALITY, RELIABILITY, MAINTAINABILITY,
VALUE ENGINEERING & ASSOCIATED
SPECIFICATIONS

1. MIL-Q-9858 A   Quality Program Requirements

This specification requires an effective and
economical quality program throughout all
areas of contract performance: Design Development, Fabrication, processing, assembly, inspection, test, maintenance, packaging, shipping,
storage and site installation.

2. MIL-I-45208 A   Inspection Systems Requirements

This specification establishes requirements
for contractors' inspection systems. Contains
fewer requirements than MIL-Q-9858. The contractor may use at his option, the requirements
of MIL-Q-9858, in whole or in part, whenever
this specification is specified providing no
increase in price or fee is involved. Forms
a part of MIL-Q-9858.

3. MIL-C-45662 A   Calibration Systems Requirements

This specification provides for the establishment and maintenance of a calibration system
to control the accuracy of measuring and test
equipment. Forms part of MIL-Q-9858 and MIL-I-
45208.

4. MIL-Q-21549   Quality Assurance Program Requirements for Fleet
   (NOrd)   Ballistic Missile Weapon System Contractors.

This specification shall apply to all Fleet
Ballistic Missile prime contractors, subcontractors or suppliers concerned with end-use
quality of materials or services. A very demanding document.

5. NPC 200-2   Quality Program Provisions for Space System
   (NASA)   Contractors

This specification sets forth general requirements for quality programs from design conception

to the delivery of articles. A very detailed document in every respect from design to the field.

6. NPC 200-3 <u>Inspection System Provisions for Suppliers of Space Materials, Parts, Components & Services</u>

This specification sets forth the minimum requirements for A Suppliers' Inspection System.

7. NPC 200-4 <u>Soldering of Electrical Connections(High Reliability)</u>

This document is very specific & demanding with regard to Soldering, Tools, Methods, Certification, etc.

8. DOD 2000 <u>Soldering & Workmanship Requirements</u>

Standard Soldering Requirements & Specific Workmanship are covered in all initial paragraphs; Paragraph 5 and up is invoked only by contract & this includes Training, Certification, Hand & Wave & Solder Analysis plus very demanding Workmanship.

9. ANSI/ASQC <u>Quality Systems Q91, Q92, Q93 of 1987 and ISO 9001,</u>
   Q-90 SERIES <u>9002, 9003 of 1987</u>(These documents are alike)
     and
   ISO 9000   Q91/9001 Covers Q.A. in Design/Develop./Prod./Instal/Serv
   SERIES   Q92/9002 Covers Q.A. in Production & Installation
               Q93/9003 Covers Q.A. in Final Inspection & Test
               Represent "Functional or Organizational Capability"
               suitable for two-party contractual purposes.

10. MIL-Std-785 <u>Requirements for Reliability Program(Systems & Equip.)</u>

Requires the most Comprehensive Reliability Program yet specified. Requires vendor

reliability programs, training programs, human
engineering, statistical models, effects of
storage analysis, parts improvement programs,
failure effects anlaysis, parameter drift
analysis, safety margins, manufacturing
reliability controls, design reviews, development testing, reliability demonstrations, contract reliability compliance considerations,
control of manufacturing processes, extensive
failure analysis reporting program, longevity
testing, a major quality assurance program –
and much more.

11. MIL-R-27542   Reliability Program Requirements for Aerospace
    (USAF)   Systems, Subsystems and Equipment

Requires same as MIL-STD-785 and has been
superseded by 785.

12. MIL-STD-441   Reliability of Military Electronic Equipment

Requires a comprehensive design review, pre-
diction, worst case analysis, and a failure
reporting and analysis. Also covers other
areas such as maintainability and quality
assurance. Has been superseded by MIL-STD-785.

13. MIL-M-26512   Maintainability Program Requirements for Aero
    (USAF)   Space Systems and Equipment

Requires comprehensive maintainability program
and program plan for contractors. Requires
quantitive and qualitative characteristics and
principles; planned maintenance and support;
Time (MTTR, etc.); Rate (M.M. hrs., etc.); com-
plexity (people/skills); costs of maintenance;
accuracy; design principles per AFSC Manuals
80-1, 80-3, etc., with reliability and quality.

14. MIL-V-38352   Value Engineering Program Requirements

Establishes the minimum requirements for a contractor's value engineering program, including organization, control, program elements, etc.

15. Handbook H-111   Value Engineering

Covers value engineering from definitions to program control, including methodology, criteria for applying value engineering, management review and action, organization, training, and motivation.

16. MIL-STD-105   Sampling Procedures and Tables for Inspection by Attributes

Internationally recognized in planning and conducting sampling on most every type of product, etc.

17. MIL-STD-414   Sampling Procedures and Tables for Inspection by Variables for Percent Defective

Establishes sampling plans and procedures for inspection by variables for use in government procurement, supply, storage, and maintenance inspection operations.

18. MIL-STD-1235   Single and Multilevel Continuous Sampling Procedures and Tables for Inspection by Attributes

Establishes sampling plans and procedures for use in determination of acceptability of products in procurement, supply, storage, and maintenance inspection operations. Contains tables and graphs for several continuous sampling plans, single and multilevel. Instructions for use are included.

19. MIL-STD-7   <u>Types and Definitions of Engineering Drawings</u>

Defines the types of engineering drawings prepared by the Army, Navy, and Air Force, and by contractors. The drawings reveal engineering information used for construction, evaluation, inspection, identification, maintenance, and manufacture.

20. MIL-STD-8   <u>Dimensioning and Tolerancing</u>

Establishes the rules, principles, and methods of dimensioning and tolerancing applicable to defining geometric characteristics of objects delineated on drawings.

21. MIL-STD-10
    and
    ASA B-46.1   <u>Surface Roughness, Waviness and Lay</u>

Establishes uniform system for identification and specification of the geometric irregularities of the surfaces of solid materials. Provides symbols and numerical classifications as a means for accurately expressing surface roughness, waviness, and lay requirements on drawings, in specifications or verbally.

22. MIL-STD-107   <u>Preparation and Handling of Industrial Production Equipment for Shipment and Storage</u>

Establishes requirements, guidance, and approved methods for disassembly, inspection, cleaning, preserving, packaging, packing, marking, blocking, bracing, skidding, and recording prior to shipment or storage and maintenance and surveillance of production equipment during storage.

23. MIL-STD-109   <u>Quality Assurance Terms and Definitions</u>

Designates quality assurance terms and definitions consistent with present assurance concepts held by the Government. Provides standardized interpretation of quality assurance terms and definitions as applied to determination of product quality.

24. MIL–HDBK-141   Optical Design

Provides engineering, management, and personnel
with introduction to optical theory and treats
to an advanced level the fundamentals and
principles of optical design. Covers full
range of subjects encountered in the optical
field.

25. MIL-STD-150   Photographic Lenses

Establishes uniform definitions, nomenclature,
classification of defects, methods of testing
and measurements pertaining to the field of
photographic lenses.

26. MIL-STD-1241   Optical Terms and Definitions

Establishes definitions for the words, terms,
and expressions peculiar to the general field
of optics.

27. MIL-STD-120   Gage Inspection

Provides correlated technical information
applicable to the inspection of gages, special
tools, and measuring devices. Covers nomenclature, tolerances and fits, measuring tools,
equipment, gages, and methods of measurement
and inspection.

28. MIL-STD-202   Test Methods for Electronic and Electrical
Component Parts

Establishes uniform methods for testing electronic and electrical component parts, including basic environmental tests to determine
resistance to deleterious effects of natural
elements and conditions surrounding military
operations, and physical and electrical tests.

29. MIL-E-16400   <u>Military Specification Electronic Equipment, Naval Ship and Shore: General Specification</u>

Covers the general requirements applicable to the design and construction of electronic equipment and associated and auxiliary electronic apparatus furnished as part of a complete system intended for Naval ship or shore applications. The intent of this specification is to set forth the ambient conditions within which equipment must operate satisfactorily and reliably; the general material, the process for selection and application of parts, and to detail the means by which equipment as a whole will be tested to determine whether it will so operate.

30. MIL-STD-129   <u>Marking for Shipment and Storage</u>

Provides the requirements for the uniform marking of military supplies and equipment for shipment and storage. Covers military standard requisitioning and issue procedures (MILSTRIP) and military standard transportation and movement procedures (MILSTAMP) for movement and processing.

31. MIL-STD-130   <u>Military Standard Indentification Marking of US Military Property</u>

Establishes the item marking requirements for identification purposes as required in stocking and replacing of parts, sub-assemblies, assemblies, units, sets, and all other items of military property required by the Department of Defense.

ZERO DEFECTS

it's up to you to

DO IT RIGHT THE FIRST TIME !

People are conditioned throughout their lives to accept the fact that they are not perfect and will, therefore, make mistakes. By the time they seek an industrial career, this belief is firmly rooted. It becomes normal to say, "People are human and humans make mistakes. Nothing can ever be perfect as long as people take part in it." And so it goes . . . .

WHAT IS A ZERO-DEFECTS PROGRAM ?

Zero-Defects is a challenge. It is a challenge for you to create an honest desire to produce defect-free work whether you are an operator, an inspector, an engineer, a typist, or any person employed by your company. It's a challenge to produce perfect work by preventing defects instead of detecting them. You're working in an era that demands Zero-Defects.

In many organizations there is little concern over a small percentage of errors. However, you at all times must set a strict standard of perfection. A standard we achieve too often by defect detection. Sometimes the cost of reaching this goal comes too high.

For instance, you should never let a product leave the plant unless it checks out perfect in all phases. Defects cause rescheduling, rework, reinspecting and retesting, all of which are costly and time consuming. In many cases, there is an opportunity to correct defects. In many more cases, you do not get that second chance.

But, a Quality reputation is not built on a defect correction program. It's the Quality that's "built in" from drawing board to the customer that counts. As one bad apple can spoil all the others in a barrel, so, a few security slips, a minor soldering error, some typing mistakes and other seemingly small defects can spoil all the good work done by the vast majority of the people. For this reason, we must strive to achieve perfection in every single action . . . to prevent errors by carefully watching each and every detail . . . to prevent mistakes before they happen rather than having them corrected after they occur.

Some industries accept a small percentage of dissatisfied customers as a part of life. We, however, are striving for a 100% customer satisfaction goal. It's not an easy task. Nothing worthwhile and really important ever is.

Hence, Zero-Defects is more than a challenge. It must be our way of life. Zero-Defects means "Do it right the first time."

HOW DOES ZERO-DEFECTS AFFECT YOU?

Zero-Defects re-emphasizes the need to do your best at all times. Your performance, no matter what your job, reflects on the reputation of each of us . . . one small error can erase the hard work of your fellow employees.

Must you make a certain number of mistakes in everything you do? Like cashing your pay check, for instance. If 5% of your work on the job is defective, does this mean you are short changed on 5% of your money transactions each year? Will you forget to pay your income tax 5% of the time? Will you go home to the wrong house several times each month?

Errors are frequently an indication of the importance a man places on specific things. He is more careful about one thing than another. He feels that it is all right to make mistakes in his work, but not permissable to defraud the government. In short, he has developed a dual attitude. In some things he is willing to accept errors, in others the amount of defects must be zero.

WHY DEFECTS?

Mistakes are basically caused by two things:

> Lack of knowledge
> Lack of attention

Knowledge can be measured and shortcomings corrected through tried and true methods.

Lack of attention is an attitude problem. It must be corrected by the person himself. A man who commits himself to watch each detail and carefully avoid errors takes a giant step toward setting a goal for Zero-Defects in all things.

WHAT IS TO BE GAINED BY A ZERO-DEFECTS MODE OF OPERATION?

The rewards are many. But, the most important reward of all is the knowledge that you have contributed to the growth of the company and the reputation of your section, department or division. By so doing, you have made yourself a more valuable asset to your family, your company, your community, and your country.

ZERO DEFECTS = High Quality and Reliability at Lowest Cost.

Join the Zero Defect Team . . . . .

"DO WHATEVER YOU DO RIGHT THE FIRST TIME"

TRAINING, QUALIFICATION & CERTIFICATION OF PERSONNEL

The art and science of producing reliable parts and equipment using "Special Processes" requires well-trained, qualified and motivated people who are capable of and can demonstrate their ability prior to producing a product for sale to commercial, military or space oriented customers.

It doesn't cost any more money to produce reliable, quality products than it does to produce questionable, poor quality products. In fact, it will cost more in the end to produce the poor quality because of the rework, test failures, trouble shooting time, misuse and abuse of tools and possible destruction of equipment. We all must agree and acknowledge the basic fact that the cost effective way to operate is to DO IT RIGHT THE FIRST TIME !

All personnel assume a role of responsibility with regard to the quality and reliability of the finished product. Only through qualification and certification can quality of workmanship be built into the product. The design engineer must design the quality and reliability into the product. The inspector cannot "inspect" quality and reliability into the product. All personnel must understand what he or she is doing, and by this we mean they must comprehend and know their job function, how it's done, and all the proper tools, methods and techniques employed to DO THE JOB RIGHT THE FIRST TIME.

A concerted qualification and certification program, tailored to the operation of the company will provide the company with qualified personnel who have demonstrated their ability to build products to company standards and specifications, thus giving management a high degree of confidence in each person as an individual who will make up the first line quality team for . . . .

QUALITY - RELIABILITY - VALUE - INTEGRITY

TYPICAL CERTIFICATION CARD

10

TEN POINTS OF QUALITY

1. <u>QUALITY IS THE "EXTRA SOMETHING"</u>, the distinction that causes a product to stand out in its field.

2. <u>QUALITY REFLECTS ITS MAKER</u>. Pride of workmanship, experience, care shown in every product.

3. <u>QUALITY IS A BRAND NAME</u>. Products are bought when customers recognize and trust a familiar trade mark or company.

4. <u>QUALITY IS A BUILDER</u>. Century-old companies gained and held success by providing good products and good service for its customers.

5. <u>QUALITY IS AN ATTITUDE</u>. The personal attitudes of employees toward their job, their company, their products will determine the degree of product quality.

6. <u>QUALITY IS MANY PEOPLE</u>. Every person is responsible for doing their job right the first time.

7. <u>QUALITY IS PLANNED</u>. As a company we must plan to make a top-quality product at a reasonable price and deliver on schedule.

8. <u>QUALITY, RELIABILITY AND VALUE</u> of our products are always remembered by the customer long after the cost is forgotten.

9. <u>QUALITY, PRICE AND DELIVERY</u> (or P. D. Q.) are achieved when the job is done "Right the First Time."

10. <u>TOTAL QUALITY ASSURANCE IS THE TEAMWORK APPROACH TO SATISFIED CUSTOMERS.</u>

VENDOR QUALITY ASSURANCE

Adequate control of Procurement Sources is necessary for all types of industries in order to assure Price - Delivery - Quality and Value.

Quality Assurance Requirements are invoked by the many documents as previously described and/or may be stipulated in the contract and terms portion of the actual Purchase Order or may be covered by individual company Specifications to which the vendor must comply. Most of the major prime contractors invoke their own Specifications which gives them specific control over their vendors.

Vendor surveys and management evaluations are used to develope a confidence level in the vendor's ability to provide quality products, on time at the stated cost. The survey must be made by qualified, knowledgable personnel who have been adequately trained in evaluating vendors, quality systems and management. The survey is an "early warning system" and should be used as a guide in evaluating the vendors' capabilities.

A Typical Survey and Evaluation form has been provided for your information and guidance.

VENDOR QUALITY ASSURANCE
EVALUATION/SURVEY/AUDIT

PROCEDURE: Complete this Vendor Evaluation and Management Survey form in detail as appropriate to the ability and capability of the Supplier you are evaluating. Not every Supplier will have ALL of the Functions covered: Example is a Distributor, Welding Shop, Paint Shop, etc. In these cases, the experienced Evaluator will only look for specific prevention, control and assurance procedures and functions. This form will be used by Purchasing and Quality Assurance Management. This will become a Permanent Management Record. This is a "Confidential Record" and is NOT to be disclosed to the Vendor or any person(s) external to THIS COMPANY!

Date:_____Survey No.:_____Conducted By: _____
Company Name:_____ ( )
Address:_____
Zip:_____Phone: _____
Division or Subsidiary of: _____
President: _____( ) VP/GM_____( )
Chief Eng.: _____( ) Dir.Mfg._____( )
Dir.Qual.: _____( ) Dir.Purch.:_____( )
Comments:_____

Principal Product/Process/Service: _____
Total Employees:_____ Mfg._____ Qual.____ Eng.____ Avg.Time W/C____
Training Program? Yes____ No____ Operators Certified? Yes ____ No ____
Familiar With: OSHA Yes____ No____ Equal Employment Opportunity? _____
Quality Control Manual Available? Yes____ No____ Revision Date: _____
If Not Available Now, When? Give Specific Date: _____
Names of Some Major Customers:_____

Approval of Quality Manual by Customers? Yes____ No____ Letters Avail?____
Quality Manual Per: ANSI/ASQC-90 M-1____ MIL-Q-9858____ MIL-I-45208____
ANSI-Z-1.8_____ Canada: 1015/1016_____MIL-STD-45662____ CSA___ISO-9000____
Facility Equipment List Available: Yes____ No____ If No, When? _____
Can We/Source Inspect at your Facility? Yes____ No ____
Product(s) U/L?/CSA? Yes_____ No____ Famil./Baldrige Award? Yes___ No___

SURVEY RATING

Procedures:_____Suppliers/Vendors:_____Dwg./Chg.Control____Insp.Eqp.____
Discrepant Mtl. Control_____Recv.Insp.____In-Proc.Insp.____Final____
Pack/Ship Insp.____Stock Control____Environ. Test____Housekeeping____
Sample Insp.____Process Control____Rel/Design Rev.____Mgt./Organ.____
Comments:_____
Avg.Rating:____Approved:_____Conditional Apprvd.____Not Apprvd.____
Vendor Capability Rating Meanings: 0-1 Little or No System (Unacceptable)
 1-2 System Inoperative; 2-3 Major Problems; 3-4 Minor Prob. to Accepted
     (Unacceptable-   (Conditional appvl.-   (Conditional to Approved
     willing to correct   will correct problems)   Status)
     deficiencies)
Management People Contacted: _____

Comments:_____
_____
_____
_____

NOTE: Number in ( ) indicates years in business or time with company.

COMPANY NAME: _____

1. GENERAL INFORMATION

1. Have Union____Name_____
   Contract Date_____
2. Have Quality Cost Program?_____
   Mgt. Review?_____When_____
3. Gov't appvd. purch. dept. _____
   Agency_____
4. Purch.procedures in effect?_____

5. Last year's gross sales_____
6. Financial Statement available?____
7. Security regulations required ?___
   Clearance level
8. Number of buildings_____
   Condition
9. Growth & Expansion Plans?_____
   _____
10. Customer Dock to Stock Program
    Yes_____No_____

11. J-I-T Program? Yes__ No__

2. PROCEDURES

Y N R
__ __ __ 1. Maintain written Quality
             and Process Control Procedures?
__ __ __ 2. Maintain Inspection & test
             instructions?
__ __ __ 3. Maintain workmanship
             Stds?
__ __ __ 4. Compliance to established
             procedures, stds.,etc?
__ __ __ 5. TQM System?
_____Total   Average_____

3. CONTROL OF SUPPLIES & VENDORS
Y N R
__ __ __ 1. Qualify suppliers' parts
             prior to use?
__ __ __ 2. Maintain qualified pro-
             ducts list, (Gov't and/
             or company)?
__ __ __ 3. Review suppliers' Quality
             Control System-Documented?
__ __ __ 4. Quality requirements in-
             cluded in purchase docu-
             ments (certification;
             test reports; quality
             system, etc.?)
__ __ __ 5. Purchase document review
             or audit?
__ __ __ 6. Maintain vendor analysis/
             rating program?
__ __ __ 7. Are vendor history records
             used in making procurement
             decisions?

_____ Total   Average _____

4. DRAWING AND CHANGE CONTROL

Y N R
__ __ __ 1. Specific Responsibility
             for drawing & change control
             __Q.C.,__Eng.,__Prod.
__ __ __ 2. Dwg./Spec. Location records
             maintained?
__ __ __ 3. Latest dwgs./specification
             in use?
__ __ __ 4. Q.C. responsible for change
             effectivity verification?
__ __ __ 5. Program for processing
             changes to suppliers?
__ __ __ 6. Customers notified of signifi-
             cant proprietary changes?
__ __ __ 7. Applicable company specs.
             available?
__ __ __ 8. Applicable customer specs.
             available? Military, etc.

_____   Total   Average _____

5. INSPECTION AND TEST EQUIPMENT
   CALIBRATION

Y N R
__ __ __ 1. Scheduled elec./mechanical
             calibrations?
__ __ __ 2. Effective recall system?
__ __ __ 3. Standards traceable to NIST
             Through_____
__ __ __ 4. All equipment identified
             and currently calibrated?
__ __ __ 5. Tooling used for acceptance
             included in calibration
             system?
__ __ __ 6. Employee-owned acceptance
             tools controlled?
__ __ __ 7. Primary stds.isolated to
             prevent unauthorized usage

_____Total   Average _____

14

6. MATERIAL REVIEW, CONTROL OF DISCREPANT MATERIAL

Y N R
___ ___ ___ 1. Maintain MRB? Members:
___ ___ ___ 2. Discrepant material physically identified? How?
___ ___ ___ 3. Discrepant material removed from production flow?
___ ___ ___ 4. Maintain records of all MRB action?
___ ___ ___ 5. Documented corrective action system?

_____Total   Average_____

7. RECEIVING INSPECTION

Y N R
___ ___ ___ 1. Material identified & segregated to preclude unauthorized usage?
___ ___ ___ 2. Use inspection checklists?
___ ___ ___ 3. Insp./test equip.adequate?
___ ___ ___ 4. Maintain vendor history records?
___ ___ ___ 5. Dwgs/specs/PO's available?
___ ___ ___ 6. Raw material & special process certifications available?
___ ___ ___ 7. Scheduled certification verification?
___ ___ ___ 8. Static Free Program?
_____Total   Average_____

8. IN-PROCESS INSPECTION/TEST

Y N R
___ ___ ___ 1. Perform first article insp?
___ ___ ___ 2. Performed or verified by Q.C.
___ ___ ___ 3. Instructions/checklists used?
___ ___ ___ 4. Results recorded on units, travelers or other identifying documents?
___ ___ ___ 5. Inspection status readily evident?
___ ___ ___ 6. Inspection/test equip.adequate?
___ ___ ___ 7. Inspection areas adequate?
___ ___ ___ 8. Static Free Program?
_____Total   Average_____

9. FINAL INSPECTION/TEST

Y N R
___ ___ ___ 1. Performed or witnessed by Q.C.
___ ___ ___ 2. Results recorded on units, travelers, or other identifying documents?
___ ___ ___ 3. Records retained?
___ ___ ___ 4. Instructions/checklists used?
___ ___ ___ 5. Inspection/test equip.adequate?
___ ___ ___ 6. Inspection areas adequate?
___ ___ ___ 7. Static Free Program?

_____Total   Average_____

10. PACK/PACKAGING AND SHIPPING INSPECTION

Y N R
___ ___ ___ 1. Instructions/checklist used?
___ ___ ___ 2. System to verify final acceptance?
___ ___ ___ 3. System to safeguard quality between final acceptance and shipping?
___ ___ ___ 4. Packaging test conducted, when required?
___ ___ ___ 5. Static Free Program?
_____Total   Average_____

11. STOCK CONTROL

Y N R
___ ___ ___ 1. Stock readily identified?
___ ___ ___ 2. Material traceable to receiving inspection records?
___ ___ ___ 3. Facilities & Space adequate?
___ ___ ___ 4. Stock rotation plan (first in-first out)
___ ___ ___ 5. Static Free Program?
_____Total   Average_____

15

12. ENVIRONMENTAL TEST FACILITIES

Y N R
___ ___ ___ 1. Facilities available?
              Types:
              _____
              _____
___ ___ ___ 2. Test result documentation
              available?
___ ___ ___ 3. Equip. included in cali-
              bration system?
___ ___ ___ 4. All equip. identified,
              currently calibrated &
              maintained?

_____ Total Average

13. HOUSEKEEPING/MAT'L HANDLING/ FACILITIES

Y N R
___ ___ ___ 1. Materials, supplies & work
              arranged in neat manner?
___ ___ ___ 2. Work & Storage areas clean
              and maintained?
___ ___ ___ 3. General appearance & condition
              of facility; good, neat, clean?
              COMMENTS:_____
              _____
___ ___ ___ 4. **Safety Program Oper.?**

_____ Total Average _____

14. SAMPLE INSPECTION

Y N R
___ ___ ___ 1. Sample plan used? Describe
              Where?_____
              ___Receiving ___ In-Process
              ___Final
___ ___ ___ 2. Statistically valid?
___ ___ ___ 3. Sampling schedules posted in
              inspection areas?
___ ___ ___ 4. Appropriate records available?

_____ Total Average _____

15. PROCESS CONTROL

Y N R
___ ___ ___ 1. Maintain chemical/physical
              Lab: Inside___ Outside___
___ ___ ___ 2. Systematic tests/analysis
              performed?
___ ___ ___ 3. Records of test/analysis
              results available?
___ ___ ___ 4. Applicable personnel and
              equipment certified/qualified:
              i.e., welding, etc.
___ ___ ___ 5. Statistical Process Control
              Program?
___ ___ ___ 6. Static Free Program?
_____ Total Average_____

16. RELIABILITY/DESIGN REVIEW

Y N R
___ ___ ___ 1. Separate Rel. Function?
___ ___ ___ 2. Separate Eng. Organization?
___ ___ ___ 3. Evaluation of Co. Products
              made?
              By Whom?_____
              When?_____All?_____
              By Contract_____
___ ___ ___ 4. Documented Rel. & Design
              Rev. policies and procedures?
___ ___ ___ 5. Failure Analysis Conducted?
              By Whom?_____
___ ___ ___ 6. Experience in Rel/D.R.?
___ ___ ___ 7. Field Performance Data Avail?
              How obtained?_____

_____ Total Average_____

SPECIAL PROCESSES

PROCESS	IN PLANT	SUB CONT.	CONTROL	SUBCONTRACTOR	COMMENT
P/C Boards	____	____	____	_____	____
Soldering(H)(W)	____	____	____	_____	____
Lugging	____	____	____	_____	____
Wire-Wrap	____	____	____	_____	____
Heat Treat	____	____	____	_____	____
Plating (Type)	____	____	____	_____	____
Chem.Film(Type)	____	____	____	_____	____
Painting	____	____	____	_____	____
Ultrasonic Clean	____	____	____	_____	____
X-Ray, Zyglo, Magnflux	____	____	____	_____	____
Welding (Type)	____	____	____	_____	____
Printing-Screening	____	____	____	_____	____
Metal Forming	____	____	____	_____	____
Extruding	____	____	____	_____	____
Turnkey Assy & Test	____	____	____	_____	____
Statistical P.C.Prog.	____	____	____	_____	____
Cables/Harness	____	____	____	_____	____
Semi-Cond./I.C.	____	____	____	_____	____
SMT;_____	____	____	____	_____	____

Others
ADDITIONAL COMMENTS: _____

_____

_____

_____

_____

_____

_____

_____

_____

Note: Add more sheets for notes which will assist you in assessing the vendor in a
      fair and just manner. Stop, look, listen and write it down so you can
      remember.

16.1

NOTES:

GENERAL INSPECTION CHECKLIST

Visual–Mechanical Inspection Shall be Accomplished under 4 Power Magnification.
Electrical Test Equipment Must Be in Calibration.

TYPICAL ITEMS

1. No loose hardware, parts, sub–assemblies, etc.

2. Correct screws, nuts, washers & related hardware, etc. Per Print.

3. Name plates, markings & decals shall be smooth. Stamping, screening, etc. shall be intact, legible & ink shall be fungus proof and pass the tape test.

4. Parts shall be free from scratches, fractures, dents, burrs, etc.

5. Painted surfaces per spec. & free from scratches, chips, blisters, etc.

6. All soldered connections per standard using approved tools only.

7. All solderless terminals per standard using approved tools only.

8. Wire length & routing sufficient to allow one repair.

9. Cable & wire routing per print: Tight cable lacing; No pinched or chaffed wires; Wire & cable clear of screw threads & sharp edges.

10. Adequate protection & clamping of cables.

11. No spliced wires unless specified on drawing and/or approved by Engineering & Quality Assurance.

12. No foreign material. Units shall be clean, inside and out.

13. Route tag completed. Unit stamped whenever possible by Production and Quality.

14. Unit tested per spec. and properly stamped as appropriate.

15. Proper packing, packaging & all paper work completed.

Work and Inspect in a Planned, Systematic Manner.

Be Consistent, Be Sure, Be Satisfied.

Your Stamp Means Zero Defects

TYPICAL INSPECTION TERMS & DEFINITIONS

FAILURE
The occurrence of a defective item while the item is under test or in service.

DEFECT
A defect is a deviation of an item characteristic from specified standards. (Defects may be of major or minor importance, depending upon their nature and degree).

MAJOR DEFECT
A major defect is one which reduces the usability of an item for its intended purpose.

MINOR DEFECT
A minor defect is one that does not materially reduce the usability of an item for its intended purpose, or is a departure from established standards having no significant bearing on the effective use or operation of the item.

DEFECTIVE
An item which is found upon examination to exhibit one or more defects.

CHARACTERISTIC
A characteristic is an inherent and measurable property of an item. Such a property may be electrical, mechanical, thermal, hydraulic, electromagnetic, or nuclear, and can be expressed as a value for stated or recognized conditions. A characteristic may also be a set of related values of a property, usually shown in graphical form.

LIFE TEST (ENDURANCE)
A test in which an item is subjected to specified stress(es) over a specified relatively long period of time or large number of operations, or both, in order to determine its durability.

ADDITIONAL TYPICAL TERMS AND DEFINITIONS

BREAKOUT

The point at which a wire or group of wires emerge from a laced portion of a wire harness or cable assembly.

BUS WIRE

The solid conductor used to permanently connect two or more terminals together. It may be either insulated or noninsulated.

CABLE

A combination of conductors insulated from one another and/or a multiple conductor.

COMPONENT

Any one of various electrical devices, such as a capacitor, resistor, transistor, tube, etc.

LACING CORD

Waxed cord/nylon material used to bind a group of wires in a bundle or designated pattern.

CRIMPING

The operation of attaching solderless terminals to a wire. It is the operation of squeezing (crimping) the terminal to the wire with the proper tool.

GROMMET

A rubber extrusion used in covering the sharp edges of a hole or chassis.

HARNESS

A group of wires formed and tied in a presentable pattern to accommodate designated components.

LUG

A metal device of various sizes, either soldered or crimped onto the end of a conductor. The lug is a mechanical means of making an electrical connection at a terminal or terminal strip.

LACING

A method of binding a group of wires in a bundle or designated pattern.

PIGTAIL

The wire attached to the shield for terminating purposes, or the conductor extending from a small component.

POINT-TO-POINT	When pertaining to wire, is that wire which takes the most direct route between terminals, provides no service loops, yet is not taut.
SERVICE LOOP	A predetermined portion of wire or conductor to facilitate maintenance and relieve tension.
TYE WRAP/SPOT TIE	A tie made with cord or plastic to temporarily or permanently tie groups of wires together at one point.
SOLDERED JOINTS	The connection of similar or dissimilar metals by applying molten solder, with no fusion of the base metals.
PRINTED CIRCUIT ASSEMBLY	A printed circuit board on which separately manufactured component parts have been added.
FLOW OR DIP SOLDERING	The process whereby assemblies are brought in contact with the surface of molten solder for the purpose of making soldered connections.
HELIARC WELDING	An electrical-arc process in which the welding zone is protected by a shield of helium or argon gas.
SPOT WELD	A type of weld used to join two overlapping plates. A pressure welding process in which the fusion is confined to a relatively small portion of the area of the lapped parts to be joined.
ARC WELDING	A non-pressure (fusion) welding process in which the welding heat is obtained from an arc either between the base metal and an electrode or between two electrodes.

(Typical types of welds are: Butt & Joint; "T" Joint; Lap Joint; Corner Joint; Edge Joint.)

DEFINITIONS OF DRAWING TERMS

REFERENCE DIMENSION (REF)

These are dimensions added to a drawing to simplify or record useful information for design or manufacturing.

TOTAL INDICATOR READING (TIR)

FULL INDICATOR READING (FIR)

This is the total hand movement of an indicator set to record the amount that a surface varies from absolute concentricity or roundness, or to record other variations.

OUT OF ROUND

It is the total deviation from a true circle. It is the total difference between the measured maximum and minimum diameters.

DATUM POINT

This applies to the base from which working diameters or dimensions may originate.

CONCENTRICITY

This is a condition where the axis of one symmetrical feature coincides with the axis of one or more other symmetrical features within a part.

ECCENTRICITY

This is a condition where the axis of a particular feature is parallel to, but offset from the axis of another feature; or where the axis of a rotating part mounted in an assembly does not coincide with the axis of the part about which it turns.

TYPICAL (TYP)

When associated with a dimension, means that the dimension applies to all features that appear to be identical in size and configuration. The tolerance stated for a dimension labeled typical also applies to each identical feature.

DEFINITIONS OF DESTRUCTIVE AND NONDESTRUCTIVE TERMS
FOR
METALS AND CASTINGS

FRACTURE TEST — Breaking metal to determine structure or physical condition by examining the fracture.

MAGNETIC ANALYSIS INSPECTION — A nondestructive method of inspection for determining the existence and extent of possible defects in ferromagnetic materials.

MAGNAFLUX TEST — A method of detecting cracks, laps, and other defects by magnetizing the parts and applying fine magnetic particles (dry or suspended in solution).

RADIOGRAPHY — A nondestructive method of internal examination in which metal or other objects are exposed to a beam of x-ray or gamma radiation.

ZYGLO INSPECTION — Metals are treated with a special dye containing water washable oil which has the power to penetrate extremely small surface cracks. The part is then illuminated with short wave-length light called "black light" causing the dye to glow with fluoresence - thus indicating the size and location of cracks or other defects.

ULTRASONIC IMAGING — Conversion of penetrating ultrasonic energy into electrical energy for presentation on a TV monitor.

INFRARED — Measurement of variations in emitted infrared radiation caused by the presence of discontinuities.

Acceptable Quality Levels (normal inspection)

Sample size code letter	Sample size	0.010		0.015		0.025		0.040		0.065		0.10		0.15		0.25		0.40		0.65		1.0		1.5		2.5		4.0		6.5		10		15		25		40		65		100		150		250		400		650		1000	
		Ac	Re	Ac	Re	Ac	Re	Ac	Re	Ac	Re	Ac	Re	Ac	Re	Ac	Re	Ac	Re	Ac	Re	Ac	Re	Ac	Re	Ac	Re	Ac	Re	Ac	Re	Ac	Re	Ac	Re	Ac	Re	Ac	Re	Ac	Re	Ac	Re	Ac	Re	Ac	Re	Ac	Re	Ac	Re
A	2	↓		↓		↓		↓		↓		↓		↓		↓		↓		↓		↓		↓		↓		↓		↓		↓		0	1	1	2	2	3	3	4	5	6	7	8	10	11	14	15	21	22	30	31
B	3	↓		↓		↓		↓		↓		↓		↓		↓		↓		↓		↓		↓		↓		↓		↓		0	1	1	2	2	3	3	4	5	6	7	8	10	11	14	15	21	22	30	31	44	45
C	5	↓		↓		↓		↓		↓		↓		↓		↓		↓		↓		↓		↓		↓		↓		0	1	1	2	2	3	3	4	5	6	7	8	10	11	14	15	21	22	30	31	44	45	↑	
D	8	↓		↓		↓		↓		↓		↓		↓		↓		↓		↓		↓		↓		↓		0	1	1	2	2	3	3	4	5	6	7	8	10	11	14	15	21	22	30	31	44	45	↑		↑	
E	13	↓		↓		↓		↓		↓		↓		↓		↓		↓		↓		↓		↓		0	1	1	2	2	3	3	4	5	6	7	8	10	11	14	15	21	22	30	31	44	45	↑		↑		↑	
F	20	↓		↓		↓		↓		↓		↓		↓		↓		↓		↓		↓		0	1	1	2	2	3	3	4	5	6	7	8	10	11	14	15	21	22	30	31	44	45	↑		↑		↑		↑	
G	32	↓		↓		↓		↓		↓		↓		↓		↓		↓		↓		0	1	1	2	2	3	3	4	5	6	7	8	10	11	14	15	21	22	30	31	44	45	↑		↑		↑		↑		↑	
H	50	↓		↓		↓		↓		↓		↓		↓		↓		↓		0	1	1	2	2	3	3	4	5	6	7	8	10	11	14	15	21	22	30	31	44	45	↑		↑		↑		↑		↑		↑	
J	80	↓		↓		↓		↓		↓		↓		↓		↓		0	1	1	2	2	3	3	4	5	6	7	8	10	11	14	15	21	22	30	31	44	45	↑		↑		↑		↑		↑		↑		↑	
K	125	↓		↓		↓		↓		↓		↓		↓		0	1	1	2	2	3	3	4	5	6	7	8	10	11	14	15	21	22	30	31	44	45	↑		↑		↑		↑		↑		↑		↑		↑	
L	200	↓		↓		↓		↓		↓		↓		0	1	1	2	2	3	3	4	5	6	7	8	10	11	14	15	21	22	30	31	44	45	↑		↑		↑		↑		↑		↑		↑		↑		↑	
M	315	↓		↓		↓		↓		↓		0	1	1	2	2	3	3	4	5	6	7	8	10	11	14	15	21	22	30	31	44	45	↑		↑		↑		↑		↑		↑		↑		↑		↑		↑	
N	500	↓		↓		↓		↓		0	1	1	2	2	3	3	4	5	6	7	8	10	11	14	15	21	22	30	31	44	45	↑		↑		↑		↑		↑		↑		↑		↑		↑		↑		↑	
P	800	↓		↓		↓		0	1	1	2	2	3	3	4	5	6	7	8	10	11	14	15	21	22	30	31	44	45	↑		↑		↑		↑		↑		↑		↑		↑		↑		↑		↑		↑	
Q	1250	↓		↓		0	1	1	2	2	3	3	4	5	6	7	8	10	11	14	15	21	22	30	31	44	45	↑		↑		↑		↑		↑		↑		↑		↑		↑		↑		↑		↑		↑	
R	2000	↓		0	1	1	2	2	3	3	4	5	6	7	8	10	11	14	15	21	22	30	31	44	45	↑		↑		↑		↑		↑		↑		↑		↑		↑		↑		↑		↑		↑		↑	

↓ = Use first sampling plan below arrow. If sample size equals, or exceeds, lot or batch size, do 100 percent inspection.

↑ = Use first sampling plan above arrow.

Ac = Acceptance number.

Re = Rejection number.

Sample size code letters — General Inspection Levels II

Code	Lot or Batch Size		Code	Lot or Batch Size		Code	Lot or Batch Size
A	2 to 8		F	91 to 150		L	3201 to 10000
B	9 to 15		G	151 to 280		M	10001 to 35000
C	16 to 25		H	281 to 500		N	35001 to 150000
D	26 to 50		J	501 to 1200		P	150001 to 500000
E	51 to 90		K	1201 to 3200		Q	500000 and over

(Extracted from MIL-STD-105)

TYPICAL
MECHANICAL INSPECTION
FOR
SPECIFIC ITEMS

1. CABLES

    a. Identify the part
    b. Visually check for damage
    c. Check for correct marking
    d. Measure cable length
    e. Continuity

2. CAPACITORS

    a. Identify the part
    b. Visually inspect for damage
    c. Physical mounting provision shall meet
       requirements
    d. Physical size shall meet the dimensions of
       the part specified
    e. Refer to ELECTRICAL TEST SPEC.

COLOR CODE

COLOR	SIG. FIGURE	MULTIPLIER	TOLERANCE PERCENT	CHARACTERISTICS
BLACK	0	1	20(M)	
BROWN	1	10		B
RED	2	100	2(G)	C
ORANGE	3	1,000		D
YELLOW	4			E
GREEN	5			F
BLUE	6			
VIOLET	7			
GRAY	8			
WHITE	9			
GOLD		x0.1	5(J)	
SILVER			10(K)	

FIXED MICA DIELECTRIC

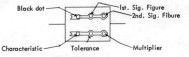

Black dot — 1st. Sig. Figure
2nd. Sig. Fibure

Characteristic — Tolerance — Multiplier

Black dot in upper left hand corner indicates MIL or JAN Mica Capacitor.

Do not read this dot when identifying capacitor.

CERAMIC CAPACITORS

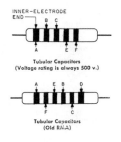

INNER—ELECTRODE END

Tubular Capacitors
(Voltage rating is always 500 v.)

Tubular Capacitors
(Old RMA)

Feed Through Capacitors

5-Dot Disc Capacitors
(EIA)
(Voltage rating is
500 v. or as marked)

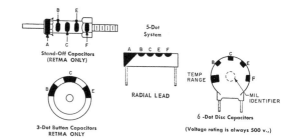

Stand-Off Capacitors
(RETMA ONLY)

5-Dot
System

RADIAL LEAD

3-Dot Button Capacitors
RETMA ONLY

TEMP
RANGE

MIL
IDENTIFIER

6-Dot Disc Capacitors

(Voltage rating is always 500 v.)

CODING FOR CERAMIC CAPACITORS

A. Temp. Coef.

B. 1st Sig. Figure

C. 2nd Sig. Figure

D. 3rd Sig. Figure

E. Multiplier

F. Tolerance

3. CHOKE COIL

    a. Identify the part
    b. Visually inspect for damage
    c. Check for correct marking
    d. Physical size shall meet the dimensions
       of the part specified
    e. Refer to ELECTRICAL TEST SPEC.

        COLOR CODING MOLDED CHOKE COILS
For small chokes, dots may be used instead of bands as above.

EXAMPLES

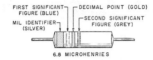

FIRST SIGNIFICANT FIGURE (BLUE)
DECIMAL POINT (GOLD)
MIL IDENTIFIER (SILVER)
SECOND SIGNIFICANT FIGURE (GREY)

6.8 MICROHENRIES

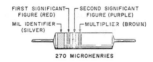

FIRST SIGNIFICANT FIGURE (RED)
SECOND SIGNIFICANT FIGURE (PURPLE)
MIL IDENTIFIER (SILVER)
MULTIPLIER (BROWN)

270 MICROHENRIES

4. CONNECTORS

    a. Identify the part
    b. Check part number stamping if applicable
    c. Visually inspect for:
       1. Damage
       2. Bent pin
       3. Long or short pins
    d. Mate with mating connector if available
    e. Check insert and/or shell is clocked as
       specified by dwg.

TYPICAL TYPES OF CONNECTORS AND THEIR POSITIONS

ALTERNATE INSERT POSITIONS

FACE VIEW OF SOCKET INSERT

NORMAL	POS. V	POS. W	POS. X	POS. Y

POS. Z	POS. E	POS. F	POS. G	POS. H

ALTERNATE SHELL POSITIONS

Typical as viewed from engaging end of shell types —11, —13 and —23. Shell types —12, —14 and —24 are exactly opposite.

NORMAL POS.	POS. X	POS. W	POS. Y

5. **DIODES**
   a. Identify the part
   b. Visually inspect for damage
   c. Check for correct marking
   d. Refer to ELECTRICAL TEST SPEC. (NOTE: Do not exceed maximum rating of unit being tested.)

The cathode end is indicated by a double-width band as the first band; or bands should be grouped toward the cathode end. The cathode end may be indicated by a single band or by the bar of the diode symbol: Cathode —▷|— Anode.

Examples   2 or 3 DIGIT TYPES OF DIODES

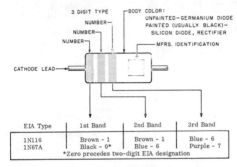

EIA Type	1st Band	2nd Band	3rd Band
1N116	Brown – 1	Brown – 1	Blue – 6
1N67A	Black – 0*	Blue – 6	Purple – 7
*Zero precedes two-digit EIA designation			

---

4 DIGIT TYPES OF DIODES

EIA Type	1st Band	2nd Band	3rd Band	4th Band	5th Band
1N1234A	Brown – 1	Red – 2	Orange – 3	Yellow – 4	Brown – A
1N1695	Brown – 1	Blue – 6	White – 9	Green – 5	Black – 0

6. **DIODE ZENER**
   a. Identify the part
   b. Visually inspect for damage
   c. Check for correct marking
   d. Refer to ELECTRICAL TEST SPEC. (NOTE: Do not exceed maximum rating of unit being tested.)

anode ▷|— cathode

7. LAMPS

    a. Identify the part
    b. Visually inspect for damage
    c. Check for correct marking if applicable
    d. Refer to ELECTRICAL TEST SPEC. (NOTE: Do not exceed Mfg. rated VOLTAGE.)

TYPICAL TYPES OF LAMPS

8. METERS

    a. Identify the part
    b. Visually inspect for damage
    c. Check dial for correct numbers
    d. Check marking on terminals (DC meter only)
    e. Refer to ELECTRICAL TEST SPEC. (NOTE: If meter is DC it is necessary to check POLARITY before testing.)

9. POWER SUPPLY

    a. Identify the part
    b. Visually inspect for damage
    c. Check for correct marking
    d. Refer to ELECTRICAL TEST SPEC. (NOTE: Care should be taken that the input voltage meets Mfg. requirements. Maximum rated load should not be exceeded.)

10. PRINTED CIRCUIT BOARD

    a. Check dimensions of board per detail dwg.
    b. Check for warpage of base material
    c. Check insulator surface-free from metal and defects
    d. Check conductor defintion and continuity
    e. Check for discoloration and delamination
    f. Check that all printed circuitry is bonded to base material
    g. Check plating
    h. Check for damage of board and/or components
    i. Check for cleanliness
    j. Check components per list of material and assembly dwg.
    k. Check for proper orientation of polarized capacitors and/or semi-conductors
    l. Check for proper soldering
    m. Check for proper positioning of components
    n. Check maximum allowable dimensions of components extension from board
    o. Check all other requirements, if not listed above per dwgs. and/or specification
    p. Refer to ELECTRICAL TEST SPEC.

11. RESISTORS

    a. Identify the part
    b. Visually inspect for damage
    c. Check for correct marking
    d. Check for correct wattage
    e. Refer to ELECTRICAL TEST SPEC.

An easy way to remember the color code

Better	Black	0
Be	Brown	1
Right	Red	2
Or	Orange	3
Your	Yellow	4
Great	Green	5
Big	Blue	6
Venture	Violet	7
Goes	Gray	8
Wrong	White	9

COLOR CODING FOR FIXED RESISTORS

COLOR	FIRST FIGURE	SECOND FIGURE	NUMBER OF ZERO	TOLERANCE
Black	0	0	- - - -	
Brown	1	1	0	
Red	2	2	00	
Orange	3	3	000	
Yellow	4	4	0,000	
Green	5	5	00,000	
Blue	6	6	000,000	
Violet	7	7	0,000,000	
Gray	8	8	00,000,000	
White	9	9	000,000,000	
Gold			x .1	5%
Silver			x .01	10%
No Color				20%

12. RELAYS
    a. Identify the part
    b. Visually inspect for damage
    c. Check for correct marking
    d. Mate with socket, when applicable
    e. Check dimensions of relay per detail dwg.
    f. Refer to ELECTRICAL TEST SPEC.

13. SOCKETS
    a. Identify the part
    b. Visually inspect for damage
    c. Mate with mating part if available
    d. Check for plating and/or discoloration
    e. Check dimensions per detail dwg.

14. SEMI-CONDUCTORS
    a. Identify the part
    b. Visually inspect for damage
    c. Check for correct marking
    d. Check dimensions per detail dwg.
    e. Refer to ELECTRICAL TEST SPEC. (NOTE: Do not exceed maximum rating of unit being tested)

TYPICAL TYPES OF SEMI-CONDUCTORS AND SYMBOLS

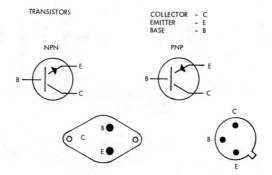

TRANSISTORS

COLLECTOR - C
EMITTER - E
BASE - B

NPN

PNP

UNIJUNCTION

BASE ONE – BI
BASE TWO – B2
EMITTER – E

SILICON CONTROLLED RECTIFIER (SCR)

CATHODE – C
ANODE – A
GATE – G

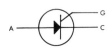

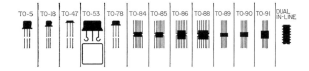

| TO-5 | TO-18 | TO-47 | TO-53 | TO-78 | TO-84 | TO-85 | TO-86 | TO-88 | TO-89 | TO-90 | TO-91 | DUAL IN-LINE |

15. TRANSFORMERS
    a. Identify the part
    b. Visually inspect for damage
    c. Check for correct marking
    d. Refer to ELECTRICAL TEST SPEC.

16. TUBES
    a. Identify the part
    b. Visually inspect for damage
    c. Check for correct marking
    d. Refer to ELECTRICAL TEST SPEC.

TYPICAL FASTENERS

TORQUE & FINISH DATA

American National and Unified Thread Nomenclature

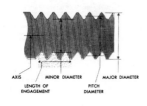

AXIS   MINOR DIAMETER   MAJOR DIAMETER

LENGTH OF   PITCH
ENGAGEMENT   DIAMETER

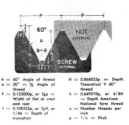

A = 60° Angle of thread
a = 30° = ½ Angle of thread
F = 0.125000p, or ⅛p = Width of flat at crest and root
f = 0.108253p, or ⅛H, or 1/6h = Depth of truncation

H = 0.866025p = Depth Theoretical V 60°
h = 0.649519p, or 6/8H = Depth American National form thread
n = Number threads per inch
p = 1/n = Pitch

SCREW THREAD MEASUREMENT

Fixed or indicating gages are used to check several varying elements which can affect interchangeability. The most important of these are:

- major diameter
- pitch diameter
- lead of thread
- minor diameter
- angle of thread

In addition, out-of-roundness and drunkenness can seriously affect product quality.

TERMS AND DEFINITIONS

Addendum   The addendum of an external thread is the distance, measured perpendicular to the axis, between the major and pitch cylinders or cones, respectively; the addendum of an internal thread is the distance, measured perpendicular to the axis, between the minor and pitch cylinders or cones, respectively.

Angle of thread   The angle included between the sides of the thread measured in an axial plane.

Complete thread   That part of the thread which has full form at both the crest and the root. When there is a chamfer at the start of the thread, not exceeding two pitches in length, it is included in the length of the complete thread.

Crest   The surface of the thread which joins the flanks of the thread and is farthest from the cylinder or cone from which the thread projects.

Crest truncation   The distance, measured perpendicular to the axis, between the sharp crest (crest apex) and the cylinder or cone which bounds the crest.

Dedendum   The dedendum of an external thread is the distance, measured perpendicular to the axis, between the pitch and minor cylinders or cones, respectively; the dedendum of an internal thread is the distance, measured perpendicular to the axis, between the major and pitch cylinders or cones, respectively.

Depth of engagement   The distance, measured perpendicular to the axis, by which the form of two mating threads overlap each other.

Drunken thread	A thread in which the advance of the thread helix is irregular.
Effective size	For an external thread, the diameter derived by adding the cumulative effects of pitch and angle errors to the pitch diameter; for an internal thread, the diameter derived by subtracting these cumulative effects from the pitch diameter.
Effective thread	The complete thread and that portion of the incomplete thread which has fully formed roots, but whose crests are not fully formed.
Flank	Either surface connecting the crest with the root, the intersection of which with an axial plane, is a straight line.
Flank angle	The angle between an individual flank and the perpendicular to the axis of the thread, measured in an axial plane. A flank angle of a symmetrical thread is commonly called the "half-angle of thread."
Form	The actual profile of a thread in an axial plane for a length of one pitch.
Helix angle	The angle made by the helix of a thread at the pitch diameter with a plane perpendicular to the axis.
Lead	The distance a screw thread advances axially in one complete turn. On a single-thread screw, lead and pitch are identical. On a double-thread, lead is twice the pitch; on a triple-thread, lead is three times the pitch.
Lead angle	On a straight thread, the angle made by the helix of the thread at the pitch line with a plane perpendicular to the axis; on a taper thread, the angle made by the conical spiral of the thread at the pitch line with the plane perpendicular to the axis at that position.
Leading flank	The flank which faces the mating thread when the thread is about to be assembled with a mating thread.
Major diameter	The largest diameter of a straight thread. On a taper thread, the largest diameter at any given plane normal to the axis.
Minor diameter	The smallest diameter of a straight thread. On a taper thread, the smallest diameter at any given plane normal to the axis.
Multiple thread	A thread in which the lead is an integral multiple of the pitch. On a double thread, the lead is equal to twice the pitch; on a triple thread, the lead is equal to three times the pitch, etc. Such threads have starting points relative to their multiple equally spaced around the circumference. For example, a double thread has two starting points 180° apart, a triple thread has three 120° apart, etc.
Pitch	The distance from a point on a thread to a corresponding point on the next thread measured parallel to the axis.
Pitch diameter	On a straight thread, the diameter of an imaginary cylinder, the surface of which would pass through the threads at such points as to make equal the width of the threads and the width of the spaces between them. On a taper thread, the diameter at a given distance from a reference plane perpendicular to the axis of an imaginary cone, the surface of which would pass through the threads at such points as to make equal the width of the threads and the width of the spaces cut by the surface of the cone.
Pitch line	The generator of the imaginary cylinder or cone specified in the definition of pitch diameter.
Root	The thread surface which joins the flanks of adjacent thread forms and is identical with or immediately adjacent to the cylinder or cone from which the thread projects.
Root truncation	The distance, measured perpendicular to the axis, between the sharp root (or root apex) and the cylinder or root which bounds the root.
Single thread	A thread in which the lead is equal to the pitch.
Size, actual	The measured value of a given dimension on an individual part.
Size, basic	The theoretical value from which the limits for a given dimension are derived by the application of allowances and tolerances.
Size, nominal	The designation used for general identification.
Thick-end thread	Threads having a thickness at the pitch diameter greater than one-half the pitch.
Thickness	Thread thickness is the distance between the flanks of the thread measured at a specified position and parallel to the axis.

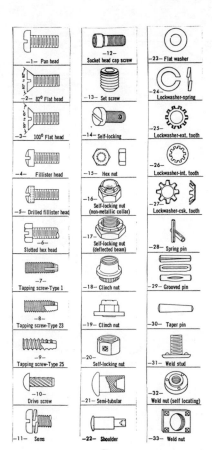

—1— Pan head

—2— 82° Flat head

—3— 100° Flat head

—4— Fillister head

—5— Drilled fillister head

—6— Slotted hex head

—7— Tapping screw-Type 1

—8— Tapping screw-Type 23

—9— Tapping screw-Type 25

—10— Drive screw

—11— Sems

—12— Socket head cap screw

—13— Set screw

—14— Self-locking

—15— Hex nut

—16— Self-locking nut (non-metallic collar)

—17— Self-locking nut (deflected beam)

—18— Clinch nut

—19— Clinch nut

—20— Self-locking nut

—21— Semi-tubular

—22— Shoulder

—23— Flat washer

—24— Lockwasher-spring

—25— Lockwasher-ext. tooth

—26— Lockwasher-int. tooth

—27— Lockwasher-csk. tooth

—28— Spring pin

—29— Grooved pin

—30— Taper pin

—31— Weld stud

—32— Weld nut (self locating)

—33— Weld nut

37

Coating or Finish for Fasteners	Used On	Corrosion Resistance
Anodizing	Aluminum	Excellent
Black oxide, blued	Steel	Indoor satisfactory, outdoor poor. Protection afforded mainly by wax or oil coatings
Black chromate	Zinc-plated or cadmium-plated steel	Added corrosion protection on cadmium and zinc-plated surfaces
Blueing	Steel	Indoor satisfactory, outdoor poor. Protection afforded mainly by wax or oil coatings
Brass plate, lacquered	Steel, usually	Fair
Bronze plate, lacquered	Steel, usually	Fair
Cadmium plate	Most metals	Very good
Clear chromate finish	Cadmium and zinc-plated parts	Very good to excellent
Dichromate	Cadmium and zinc-plated parts	Very good to excellent
Olive drab, gold or bronze chromate	Cadmium and zinc-plated parts	Very good to excellent
Chromium plate	Most metals	Good (improves with increased copper and nickel undercoats)
Copper plate	Most metals	Fair
Copper, brass, bronze misc. finishes	Most metals	Indoor, very good
Lacquering, clear or color matched	All metals	Improves corrosion resistance. Some types designed for humid or other severe applications
Lead-tin	Steel, usually	Fair to good
Bright nickel	Most metals	Indoor excellent. Outdoor good if thickness is at least 0.0005 in.
Dull nickel	Most metals	Same as bright nickel
Passivating	Stainless steel	Excellent
Phosphate Bearing Surfaces, Army 57-0-2, Type II, Class A	Steel	Good
Phosphate Rust Preventive, Army 57-0-2, Type II, Class B	Steel	Fair to good
Phosphate Paint-base Preparations, Army 57-0-2, Type II, Class C	Steel, aluminum, zinc plate	Good, after paint or lacquer applications
Colored phosphate coatings	Steel	Superior to regular phosphated and oiled surfaces
Rust preventatives	All metals	Varies with function of oil
Silver plate	All metals	Excellent
Electroplated tin	All metals	Excellent
Hot-dip tin	All metals	Excellent
Electroplated zinc	All metals	Very good
Electrogalvanized zinc	All metals	Very good
Hot-dip zinc	All metals	Very good
Hot-dip aluminum	Steel	Very good

Torque data

$T = F \times D$

where:
T = torque
F = applied force
D = distance or length of lever arm

When using this fundamental torque formula, care must be taken to measure the lever length perpendicular to the direction of the applied force.

Torque Value Conversions

Inch-Grams	Inch-Ounces	Inch-Pounds	Foot-Pounds	Centimeter-Kilograms	Meter-Kilograms
7.09	0.25				
14.17	0.5				
21.26	0.75				
28.35	1.				
113.4	4.	0.25			
226.8	8.	0.5			
453.6	16.	1.	0.08	1.11	
	32.	2.	0.17	2.35	
	48.	3.	0.25	3.46	
	64.	4.	0.33	4.56	
	80.	5.	0.42	5.81	
	96.	6.	0.5	6.91	
	112.	7.	0.58	8.02	
	128.	8.	0.67	9.26	
	144.	9.	0.75	10.37	
	160.	10.	0.83	11.48	
	192.	12.	1.	13.83	0.138
	240.	15.	1.25	17.28	0.173
	320.	20.	1.67	23.09	0.231
	384.	24.	2.	27.65	0.277
	400.	25.	2.08	28.76	0.288
	576.	36.	3.	41.48	0.415
	768.	48.	4.	55.30	0.553
	800.	50.	4.16	57.52	0.575
	960.	60.	5.	69.13	0.691
		72.	6.	82.95	0.830
		84.	7.	96.78	0.968
		96.	8.	110.61	1.106
		100.	8.33	115.17	1.152
		108.	9.	124.43	1.244
		120.	10.	138.26	1.383
		180.	15.	207.39	2.074
		200.	16.66	230.34	2.303
		240.	20.	276.51	2.765
		300.	25.	345.64	3.456
		500.	41.65	575.84	5.758
		600.	50.	691.29	6.913
		1000.	83.33	1151.68	11.517
		1200.	100.	1382.57	.13.826
		2000.	166.6	2303.37	23.034
		2400.	200.	2765.15	27.651
		5000.	416.5	5758.42	57.584
		6000.	500.	6912.86	69.129
		12000.	1000.	13825.73	138.257
		24000.	2000.	27651.46	276.515

Determination of Torque Required

A torque tool is used on a threaded fastener to control its clamping ability. The stresses induced in the body of the bolt or screw by tightening provide the clamping force. The torque applied to the head of the fastener is proportional to the load applied.

In determining the proper amount of torque, it is necessary to know the desired bolt stress. Unless this stress is fixed by the function of the assembly, it is normally based on the yield strength of the bolt material. In order to avoid premature fastener failure, induced stress is usually specified not to exceed 80% of yield strength.

For most applications, the following formula can be used to approximate the required torque:

$$T = .2dF$$

where: T = required torque

d = bolt diameter

F = bolt tension

This formula assumes the use of commercial, steel, semi-finished, regular-series nuts with rolled thread, semi-finished steel bolts on steel surfaces without lubrication. Such conditions result in a coefficient of friction of .16 at both the thread and head bearing surfaces.

The following table indicates the approximate torque requirements under the foregoing conditions, including an induced bolt stress of 40,000 lbs psi, and because of the normal variables involved, should not be taken as accurate limits.

SIZE	TORQUE	SIZE	TORQUE
0-80	16 in oz	3/8 -24	275 in lbs
1-64	24 in oz	7/16-14	390 in lbs
1-72	32 in oz	7/16-20	430 in lbs
2-56	36 in oz	1/2 -13	630 in lbs
2-64	44 in oz	1/2 -20	675 in lbs
3-48	56 in oz	9/16-12	870 in lbs
3-56	64 in oz	9/16-18	975 in lbs
4-40	5 in lbs	5/8 -11	1200 in lbs
4-48	7 in lbs	5/8 -18	1350 in lbs
5-40	8 in lbs	3/4 -10	2200 in lbs
5-44	9 in lbs	3/4 -16	2330 in lbs
6-32	10 in lbs	7/8 - 9	3450 in lbs
6-40	12 in lbs	7/8 -14	3720 in lbs
8-32	22 in lbs	1- 8	433 ft lbs
8-36	22 in lbs	1-14	475 ft lbs
10-24	29 in lbs	1-1/8 - 7	608 ft lbs
10-32	34 in lbs	1-1/8 -12	687 ft lbs
12-24	62 in lbs	1-1/4 - 7	812 ft lbs
12-28	62 in lbs	1-1/4 -12	942 ft lbs
1/4-20	70 in lbs	1-3/8 - 6	1133 ft lbs
1/4-28	80 in lbs	1-3/8 -12	1333 ft lbs
5/16-18	140 in lbs	1-1/2 - 6	1500 ft lbs
5/16-24	150 in lbs	1-1/2 -12	1666 ft lbs
3/8-16	250 in lbs		

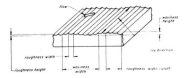

SURFACE TEXTURE DEFINITIONS

SURFACE TEXTURE

Repetitive or random deviations from the nominal surface which form the pattern of the surface. Surface texture includes roughness, waviness, lay and flaws.

ROUGHNESS

Roughness consists of the finer irregularities in the surface texture usually including those which result from the inherent action of the production process. These are considered to include traverse feed marks and other irregularities within the limits of the roughness-width cutoff.

ROUGHNESS HEIGHT

Roughness height is rated as the arithmetical average deviation expressed in microinches measured normal to the center line.

ROUGHNESS WIDTH

Roughness width is the distance parallel to the nominal surface between successive peaks or ridges which constitute the predominant pattern of the roughness. Roughness width is rated in inches.

CENTER LINE

The center line is the line about which roughness is measured and is parallel to the general direction of the profile within the limits of the roughness-width cutoff, such that the sums of the areas contained between it and those parts of the profile which lie on either side of it are equal.

ROUGHNESS-WIDTH CUTOFF

The greatest spacing of repetitive surface irregularities to be included in the measurement of average roughness height. Roughness-width cutoff is rated in inches. Standard values are 0.003″,

0.010″, 0.030″, 0.100″, 0.300″ and 1.000″. When no value is specified, the value 0.030″ is assumed.

WAVINESS

Waviness is the usually widely-spaced component of surface texture and is generally of wider spacing than the roughness-width cutoff. Waviness may result from such factors as machine or work deflections, vibration, chatter, heat treatment or warping strains. Roughness may be considered as superposed on a wavy surface.

WAVINESS HEIGHT

Waviness height is rated in inches as the peak-to-valley distance.

WAVINESS WIDTH

Waviness width is rated in inches as th spacing of successive wave peaks or successive wave valleys. When specified, the values shall be the maximum permissible.

LAY

The direction of the predominant surface pattern, ordinarily determined by the production method used.

FLAWS

Flaws are irregularities which occur at one place or at relatively infrequent or widely varying intervals in a surface. Flaws include such defects as cracks, blow holes, checks, ridges, scratches, etc. Unless otherwise specified, the effect of flaws shall not be included in the roughness height measurements.

THE SURFACE SYMBOL

The symbol used to designate surface irregularities is the checkmark with horizontal extension as shown. The point of the symbol shall be on the line indicating the surface, on the extension line or on a leader pointing to the surface. The long leg and extension shall be to the right as the drawing is read. Where roughness height only is indicated, it shall be permissible to omit the horizontal extension.

Typical applications of the symbol on a drawing

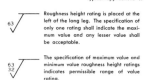

Roughness height rating is placed to the left of the long leg. The specification of only one rating shall indicate the maximum value and any lesser value shall be acceptable.

The specification of maximum value and minimum value roughness height ratings indicates permissible range of value rating.

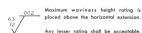

Maximum waviness height rating is placed above the horizontal extension.

Any lesser rating shall be acceptable. Maximum waviness width rating is

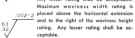

placed above the horizontal extension and to the right of the waviness height rating. Any lesser rating shall be acceptable.

Minimum requirements for contact or bearing area with a mating part or reference surface shall be indicated by a percentage value placed above the extension line as shown. Further requirements may be controlled by notes.

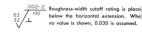

Lay designation is indicated by the lay symbol placed at the right of the long leg.

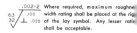

Roughness-width cutoff rating is placed below the horizontal extension. Where no value is shown, 0.030 is assumed.

Where required, maximum roughness width rating shall be placed at the right of the lay symbol. Any lesser rating shall be acceptable.

MEASURING SURFACE TEXTURE

Except for lay (which is usually observed by eye), shop instruments are available for measuring all of the surface texture characteristics defined in the standard.

Roughness and waviness rating, unless otherwise specified, shall apply in a direction which gives the maximum reading; normally across the lay.

Roughness-width cutoff is an instrument characteristic, rather than a surface-texture characteristic, although its amount is either specified or implied by the surface symbol. With continuously-averaging stylus-type instruments for measuring average roughness height, the length of trace (traversing length) used for any given measurement shall be not less than 20 times the roughness-width cutoff value. However, it is not necessary for the traversing length to be traversed continuously in one direction provided that the time required to reverse the direction of trace is short compared to the time the tracer is in motion. For this type of operation, the minimum length of travel shall be not less than 5 times the roughness-width cutoff.

In addition, if a single skid is employed to provide a reference surface and to support the tracer, it shall preferably have a radius in the direction of race of at least 50 times the roughness width cutoff. If two operative skids are used, their radii shall be not less than 8 times the roughness-width cutoff.

TYPICAL INTERVALS OF CALIBRATION

The calibration of Inspection and Test Equipment provides assurance of accuracy and acceptance of product. Calibration systems are normally patterned after Mil-C-45662 of the latest issue. The following are Guidelines for Typical Intervals of Calibration. Actual Intervals are based on the specific piece of equipment, its known Stability, Purpose and Degree of Usage.

TYPICAL INTERVALS	Maximum
Test Equipment (Scope, Bridge, Voltmeters, Etc.)	6 months
Micrometers, Calipers, Etc.	3 months
Wire Strippers (Approved Mechanical)	1 month
Wire Strippers (Automatic)	At each set up, Hourly & Random Audit.
Gage Blocks (Working)	6 months
Gage Blocks (Master)	12 months
Surface Plates	12 – 18 months

CARE AND USE OF TOOLS

Experience has shown "a craftsman is known by the tools he keeps . . . and how he keeps them". Occasionally good work may be turned out with poor tools, but this is the exception and not the rule for Quality and Reliability. All tools are made for a specific purpose. Good tools, when properly used, will assist you in acquiring job skills, continued high quality work and lasting job satisfaction.

Selection of the proper tool is very important for the job being done. Tools are costly to you and/or the company. Use them correctly. Be sure your tools are clean before you put them away. The following "DO's" and "DON'Ts" will help you get the maximum Quality, Reliability and Cost Effectiveness from your tools.

DO

1. Get all necessary tools ready before starting on an operation. PLAN AHEAD.

2. Familiarize yourself with the general operation sequence and UNDERSTAND THE JOB REQUIREMENTS.

3. Arrange tools in the order of use for production and/or inspection operations.

4. ONLY USE THE APPROVED PROPER TOOL FOR THE STATED JOB.

5. Maintain, care for and protect your tools and/or company property and equipment.

6. Repair and/or replace worn or damaged tools immediately.

7. Maintain a clean, orderly work area.

3. Only use currently Calibrated Tools and Gages.
    WORK AND INSPECT IN A PLANNED SYSTEMATIC MANNER FOR CONSISTENT QUALITY AND RELIABILITY.

DON'T

1. Start work without proper understanding of Job, Tools and Equipment.

2. Have unnecessary Tools on your workbench.

3. Stop work to get additional tools . . .

4. Misuse or abuse tools and equipment.

5. Use improper, unapproved, unmaintained tools or tools which are out of calibration.

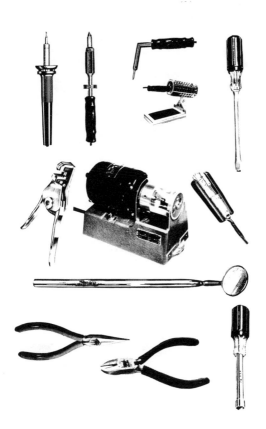

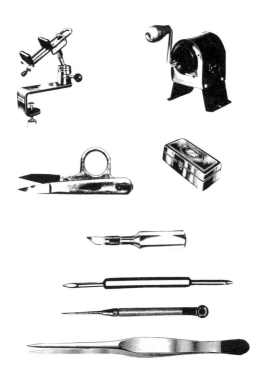

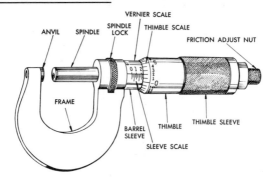

Generally, a micrometer in the hands of a skilled operator can be used to reliably measure within .0005". The instrument's inherent accuracy is usually between .00005" and .0001".

Large temperature variations can produce errors. In general, the micrometer should be near the same temperature as the test-piece, particularly for measurements below .001".

Measuring faces should be clean, free of oil, dust, lint, etc.

Measuring faces should be parallel to the test-surface, or perpendicular to the diameter of a round part.

Care should be exercised in engaging the spindle with the test-piece. Bring the measuring surfaces together slowly. A light pressure should be exerted—about 16 to 30 ounces.

Rocking the faces against the test-piece has a tendency to produce uneven wear and should be avoided.

Do not disengage the test-piece before making the reading. If the measurement cannot be viewed without removing the micrometer, use the spindle-lock (if provided) at the final setting and slide the instrument off by its frame.

Between periods of use, protect the instrument's accuracy by keeping it in its box, or stored in a safe location.

Wipe off the micrometer after use. Never use compressed air. Foreign matter may be thus forced into the spindle threads.

Regular inspection of micrometer "zero," the measuring faces, accuracy of the screw and alignment of spindle-anvil axis is recommended. Check with your distributor for detailed information on your brand.

Spindle threads should be flushed and cleaned occasionally. When dry, a few drops of high-grade light oil should be deposited using a sharp-pointed tool.

Specific length standards are available for checking the accuracy of micrometers of all capacities.

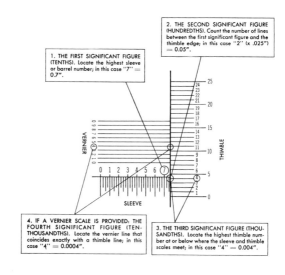

2. THE SECOND SIGNIFICANT FIGURE (HUNDREDTHS). Count the number of lines between the first significant figure and the thimble edge; in this case "2" (x .025") = 0.05".

1. THE FIRST SIGNIFICANT FIGURE (TENTHS). Locate the highest sleeve or barrel number; in this case "7" = 0.7".

4. IF A VERNIER SCALE IS PROVIDED: THE FOURTH SIGNIFICANT FIGURE (TEN-THOUSANDTHS). Locate the vernier line that coincides exactly with a thimble line; in this case "4" = 0.0004".

3. THE THIRD SIGNIFICANT FIGURE (THOUSANDTHS). Locate the highest thimble number at or below where the sleeve and thimble scales meet; in this case "4" = 0.004".

The micrometer caliper consists of a frame with an attached stud called the anvil and a movable screw which can be advanced toward, or withdrawn from this fixed anvil. The head of the screw is comparatively large and has division marks around its periphery. Most American micrometers are made with 40 threads per inch. One complete turn of the screw moves its point 1/40" (.025"). The head of the screw, or thimble, has 25 divisions on its circumference, and turning the head from one division to the next moves the point 1/1000" (.001"). In conjunction with the axial scale on its sleeve, measurements can be made to .001" over the total range of the instrument. In addition, some micrometers incorporate a vernier scale circumferentially on their sleeves permitting measurements to .0001".

PRECISION GAGE BLOCKS

A typical 81-piece gage block set contains blocks arranged in the following order:

9 Blocks - .0001" Series
.1001 .1002 .1003 .1004 .1005 .1006 .1007 .1008 .1009

49 Blocks - .001" Series

.101	.102	.103	.104	.105	.106	.107	.108	.109	.110
.111	.112	.113	.114	.115	.116	.117	.118	.119	.120
.121	.122	.123	.124	.125	.126	.127	.128	.129	.130
.131	.132	.133	.134	.135	.136	.137	.138	.139	.140
.141	.142	.143	.144	.145	.146	.147	.148	.149	

19 Blocks - .050" Series
.050 .100 .150 .200 .250 .300 .350 .400 .450 .500
.550 .600 .650 .700 .750 .800 .850 .900 .950

4 Blocks - 1.000" Series
1.000 2.000 3.000 4.000

LENGTH COMBINATIONS

Don't trust trial and error methods when assembling gage blocks into a gaging dimension. The basic rule is to select the fewest blocks that will suit the requirement.

To construct a length of 1.3275", using a typical 81-piece set, the following procedure may be used:

1. Write the desired dimension on a piece of paper.   1.3275
2. Begin the selection at the top of the gage block set.
3. Reduce the last digit of the dimension to zero by selecting a block with a 5 in the fourth decimal place. In this case, the .1005 block is selected and its length is subtracted from 1.3275 to determine a remainder still to be selected. The value .1005 may be written again in an adjacent column for subsequent proof of the selection.

$$
\begin{array}{rr}
1.3275 & \\
-\ .1005 & \\
\hline
1.2270 & .1005 \\
\end{array}
$$

$$
\begin{array}{rr}
-\ .107 & .107 \\
\hline
1.1200 & \\
\end{array}
$$

4. Select a block to reduce the third decimal place to zero.

5. Select a block to reduce the second and first decimals to zero. Wherever possible, such double reductions are desirable to cut down the total number of blocks selected.

$$
\begin{array}{rr}
-\ .120 & .120 \\
\hline
1.0000 & \\
\end{array}
$$

6. Complete the selection with the 1.000 block.

$$
\begin{array}{rr}
-1.000 & +1.000 \\
\hline
0.0000 & 1.3275 \\
\end{array}
$$

There are times when the same gaging dimension must be assembled more than once from a single set of blocks. This may unavoidably increase the number of blocks required for the specific length. Assume that a second length of 1.3275" is required from the 81-piece set:

1. Write the requirement.   1.3275
2. Select two blocks to reduce the fourth decimal to zero. In this case, .1002 and .1003.

$$
\begin{array}{rr}
-\ .2005 & .1002 \\
& .1003 \\
\hline
1.1270 & \\
\end{array}
$$

3. One block may now be selected to reduce three digits.

$$
\begin{array}{rr}
-\ .127 & .127 \\
\hline
1.0000 & \\
\end{array}
$$

4. Select two blocks to form the remaining 1.000. In this case, .400 and .600.

$$
\begin{array}{rr}
-1.000 & .400 \\
& +\ .600 \\
\hline
0.0000 & 1.3275 \\
\end{array}
$$

HOW TO WRING

- Blocks that are flat, smooth, clean and free from burrs will wring readily. Be sure that the gaging surfaces are washed clean and are absolutely dry. A small scratch without burrs will not affect wringability. A deburring stone may be used to remove burrs.
- A very minute amount of oil or kerosene applied from a pin point to the wringing surfaces will aid the operation.
- Repeatedly wipe one block on a lint-free tissue and wring together until the blocks show resistance to movement.

RECTANGULAR BLOCKS

Overlap the surfaces slightly. Maintain light pressure and slide one smoothly along the other . . . until the gaging surfaces are fully mated.

SQUARE BLOCKS

Maintaining light pressure, overlap the surfaces and slide one along the other . . . past center and . . . back to proper mating position.

TEMPERATURE COMPENSATION

Gage blocks are calibrated at the international standard measuring temperature of 60°F (20°C). When measurements are conducted at this temperature between blocks and parts of dissimilar materials, no correction for different coefficients of expansion is necessary providing the components have had sufficient time to adjust to the environment.

If blocks and parts are made of the same material and are at the same temperature, accurate results are possible regardless of whether the temperature is high or low.

To determine the correction requirement when blocks and parts are dissimilar and at temperatures other than 68°F, use the following formula:

$$E = L \,(\triangle k) \,(\triangle t)$$

where: E = the measurement error in microinches
L = nominal dimension in inches
$\triangle k$ = difference of coefficients in microinches
$\triangle t$ = deviation of temperature from 68°F

TYPICAL COEFFICIENTS OF EXPANSION
in microinches per inch of length per degree F.

Hardened Tool Steel	6.4
Stainless Steel (410)	5.5
Chrome Carbide	4.5
Tungsten Carbide	3.0
Aluminum	12.8
Copper	9.4

GENERAL PRECAUTIONS

- Don't use gage block sets having greater inherent accuracy than required by the job to be done.
- Avoid touching the gaging surfaces. Blocks should never be palmed, but should be grasped lightly on the sides—preferably by insulated forceps.
- Visually check surfaces for burrs or severe scratches that may impair wringability or scratch other blocks. Anvils and surface plates also deserve attention since they may develop potentially dangerous burrs and scratches.

- Before the measuring operation, clean the blocks carefully with pure solvent and wipe dry with lint-free tissue.
- Avoid contact with chips and abrasive dust.
- Don't leave block combinations wrung together longer than necessary.
- Before storing, wipe the blocks clean and apply a thin film of rust preventive oil or grease. When not in use, the blocks should be in their case. The case should be periodically cleaned with a stiff brush to remove accumulated foreign matter.

FOR LONGER GAGE LIFE:

- Remove abrasive grit and burrs from the part to be inspected.
- Don't expect accurate measurements of parts that are overheated or obviously at a much different temperature than the gage.
- Don't rely on gaging members that are burred or nicked. Such gages need attention and should be removed from possible use.
- Learn how to properly hold and use fixed gages. Twist the gage into the work slowly to get the

proper feel. A good fit will be snug and smooth. Never force the gage into or over the test piece. Mishandling causes excessive wear.
- To help prevent rust and expansion, don't lay gages on metal surfaces or leave them exposed to direct sunlight.
- Never clean the gaging surfaces without using a rust preventive as the final step.
- Provide gage racks for storage. Felt-covered wood that is oiled occasionally is ideal.

FIXED GAGE NOMENCLATURE

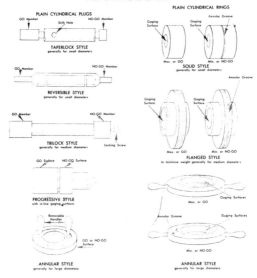

Because the exact geometry of internal and external diameters may assume a number of shapes, fixed gages are limited to assessing maximum and minimum conditions.

For example, although a GO plug may enter an i.d. and feel somewhat loose, it is doubtful whether the exact defect could be determined from the possible sources indicated here by the exaggerated illustrations.

When using snap gages, one must be particularly alert to the comparison error possible from failure to maintain perpendicularity between the gage plane and the center line of the test piece.

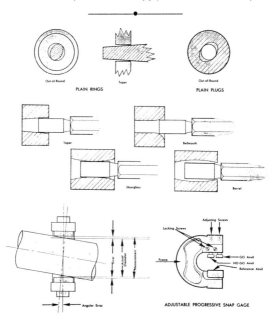

Out-of-Round

PLAIN RINGS

Taper

Out-of-Round

PLAIN PLUGS

Taper

Bellmouth

Hourglass

Barrel

Error
Actual Diameter
Measurement

Angular Error

Locking Screws
Adjusting Screws
Frame
GO Anvil
NO-GO Anvil
Reference Anvil

ADJUSTABLE PROGRESSIVE SNAP GAGE

REFERENCE SURFACES

To make precise dimensional measurements, one must have a starting point—a reference plane. Such a plane is entirely theoretical. It would be perfectly flat and have no thickness—only length and width. Therefore, it may be assumed that precise measurements are literally impossible. All one may strive for are accurate measurements—to some determinable degree.

Since a true reference plane does not exist —nor can it be created, a compromise has been made. This compromise resulted in the use of reference surfaces, and surface plates are the most common reference surfaces.

A reference surface must possess these important characteristics:
- Sufficient strength and rigidity to support the test piece.
- Sufficient and known accuracy for the measurements required.

Most surface plates are made of cast iron or natural granite. Each has certain merits.

Cast iron plates:
- Usually weigh less per relative size.

- Are not likely to chip or fracture.
- Provide acceptance for magnetic fixtures anywhere on their surface.
- Can provide a degree of wringability.

Granite plates:
- Are noncorrosive and require less maintenance effort.
- Are not prone to contact interference since they do not burr or crater.
- Have closer flatness tolerances and are cheaper per relative size.
- Have greater thermal stability.
- Are nonmagnetic.
- Are not likely to gall or retain soft metals such as aluminum.

For optimum accuracy, plates should be supported in the same manner in which they were finished. Stands made by the plate manufacturer should be used whenever possible.

Surface plates are subject to wear as are all gaging components. A system of reasonable care, periodic evaluation and recalibration will guarantee their reliability.

SUGGESTIONS FOR CARE AND USE

- Clean the surface beore each period of use.
- Put on the plate at any one time only what is required for the measurement at hand.
- Burrs on parts should be removed, and rough castings should be elevated above the surface by parallels or other accessories.
- Use extreme care in moving test pieces and apparatus on, off and along the surface. Remember that it's an accurate plane

and its condition is an integral factor in your measurement.
- Work should be distributed so that wear is not concentrated in one area.
- Particularly avoid heavy contact with the edges to prevent nicking and chipping.
- Don't leave metal objects on iron plates longer than necessary.
- Clean the surface after use and rustproof iron plates.
- Keep the surface covered when not in use.

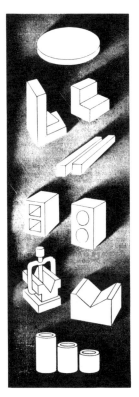

TOOLMAKER'S FLAT

When it's necessary to obtain more accurate measurements on relatively small work than are possible with ordinary surface plates, a toolmaker's flat may be used. Available in several diameters of alloy tool steel or granite, such flats can be true to less than one interference band of light (11.6 microinches).

ANGLES

Angle plates are used to provide a perpendicular surface against which the test piece may be rested or clamped. They also can be used as height blocks to elevate large work from a surface plate. Angles are available in steel or granite.

PARALLELS

Bar parallels and box parallels are used to support irregular objects above the surface plate. They are also used to support measuring apparatus in an elevated position. All working faces are parallel and square to within rigid tolerances. It's usually suggested that they be purchased in matched pairs to prevent the possibility of error if two parallels made to opposite limits were used. Parallels are made of steel or granite in a variety of sizes.

V-BLOCKS

V-blocks offer a quick, positive means of locating and holding round test pieces. Sold usually in pairs, they may be equipped with clamping devices to hold the work securely. Manufactured of steel or granite, they are also available in various sizes and grades of accuracy.

CYLINDRICAL SQUARES

Cylindrical squares provide a simple and effective means of obtaining a vertical reference line contact with a test piece in any plane. The work may be placed against the periphery and sighted, or for greater accuracy, measurements may be compared with a mechanical indicator.

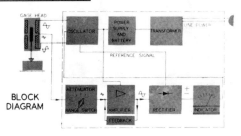

BLOCK DIAGRAM

Because of the high magnifications attainable, and the accuracy possible, electronic gaging systems assume a great reliance on environmental control. Although temperature, cleanliness and handling affect the accuracy of all measurements, the gaging techniques which may have been satisfactory for definitions in thousandths of an inch often require considerable sophistication when measuring to tenths of thousandths or millionths.

TYPICAL GAGE HEAD NOMENCLATURE:

MULTI-ANGLE GAGE HEAD

FRICTIONLESS GAGE HEAD

CARTRIDGE GAGE HEAD

HEIGHT GAGE CAUTIONS

* The stand should be placed on a surface plate whose accuracy is compatible with the measuring requirements.

* Keep the surface plate clean. Dust, oil or dirt under the height stand or test piece can affect accuracy.

* All locking knobs must be tight.

* Avoid moving the stand by grasping the column. The resulting thermal effects may expand and bend the system.

* Dress the cable from the gage head away from the area of activity to reduce the possibility of accidentally deflecting the setup.

COMMON CONSIDERATIONS

* Always select the most sensitive (highest magnification) scale of the amplifier that will entirely contain the tolerance to be checked.

* Parallax error is caused by the apparent shift of the meter pointer position when the observer views it from an angle. Always place the amplifier squarely in the line of vision so that the portion of

COMPARATOR GAGE CAUTIONS

* If the comparator stand is equipped with a removable anvil, be certain it is correctly positioned and locked in place.
* All locking knobs must be tight.
* Avoid grasping the column to move the stand.
* Allow sufficient time for thermal stabilization of the complete system and the test pieces. For optimum accuracy, place all components to be measured on a normalizing plate next to the comparator so that their temperature will match that assumed during measurement.
* Follow the manufacturer's suggestions for supporting thin test pieces, which may deform under gaging pressure.

the scale used is perpendicular to the line of sight.

* If the gaging head contact point is placed on a test piece so that the test surface is parallel to the axis of the probe, no cosine error occurs. If the contact point is at an angle to the test surface, allowances for cosine error should be made in accordance with the following table:

Angle (degrees)	Error (percent of reading)
0	0
5	+0.4
10	1.5
15	3.4
20	6.0
30	13.4

MECHANICAL INDICATING GAGES

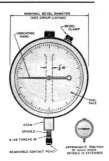
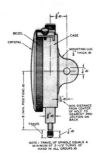

AD = OFFICIAL MONOGRAM FOR DESIGNATING PRODUCTS MADE TO AMERICAN GAGE DESIGN STANDARDS

AGD Group	Nominal Bevel Diameter (inches)		Dimension B (inches) Min. Position	Value of Smallest Graduations (in.) English	(mm.) Metric
	Above	To & Inc.			
1	1 3/8	2	1 5/8	.0001 .0005 .001	.005 .01
2	2	2 3/8	2	.00005 .0001 .0005 .001	.001 .002 .005 .01
3	2 3/8	3	2 1/8	.0001 .0005 .001	.001 .002 .005 .01
4	3	3 3/4	2 9/16	.00005 .0001 .0005 .001	.001 .002 .005 .01

Dial indicator standards extracted from Commercial Standard CS(E)119-45. Since dial numbering always indicates thousandths of an inch or hundredths of a millimeter, a quick glance can show the relative amplification of a particular gage.

DIAL GAGE CAUTIONS

- If the indicator must be used on material likely to cause excessive wear of contact points, use hard chrome, tungsten carbide or diamond tips.

- Don't tighten contact points too tight. Distortion may cause binding or sticking.

- Mount dial indicators close to the supporting fixture, securely to eliminate lost motion.

- Maintain a clean and level reference surface.

- Avoid sharp blows against a side of the contact point.

- Read the dial to the nearest indicated graduation. This will usually be found far more accurate than attempting to interpolate the fraction between divisions.

- Keep the indicator spindle clean to prevent wear and sticking.

- Don't use an indicator that has been dropped or struck until its accuracy has been thoroughly verified in accordance with manufacturer's suggestions.

- When not in use, store the indicators in a safe, dry place.

- Inspect gage accuracy on a fixed schedule governed by its use.

- Clean the dial with soap and water or an approved solvent.

- Don't drill holes in the dial case.

DIAL TYPES

BALANCED DIAL

CONTINUOUS DIAL

REVOLUTION COUNTER

AIR GAGING ADVANTAGES

- Measurement can be accomplished without physical contact between the gaging member and the test piece surface.

- The gaging member has no moving parts, which reduces the probability of wear and erratic readings.

- The cleaning effect of the air from the gaging nozzle reduces the possibility of false readings due to part contamination.

- The gaging member can be remotely positioned in relation to the indicating device.

- Gaging nozzles are relatively small and lend themselves to multiple applications for checking adjacent dimensions.

- The use of pneumatic systems requires little special skill or training.

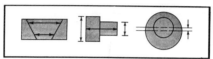

AIR GAGE CAUTIONS

- The measuring jets and any internal fixed master jets must be kept from clogging. Hence, the requirement of a clean, moisture-free air supply.

- If doubt exists about the cleanliness of a test piece surface, it is safer to clean it before gaging. Although the air has a tendency to blow away contaminants in the area of measurement, it may not do a thorough job if heavy cutting oil, grease or encrusted dirt are encountered.

- Become familiar with air gaging limitations insofar as surfaces of comparatively high roughness are concerned. Generally, conventional air gages are not recommended for surfaces whose roughness is more than 40-50 microinches AA.

- For maximum accuracy, be sure the gaging member size range and geometry match the part to be measured.

- As with all such inspection devices, allow for temperature acclimation after the system is moved to a new location.

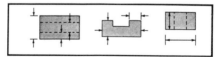

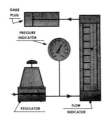

Pneumatic Gaging Air Flow System

Filtered air passes through a regulator and is controlled at an optimum pressure. It then flows up through a transparent tapered tube. An indicating float is suspended in the tube's air column and moves up and down according to the flow of air. From the top of the tube, the air moves through a hose and exhausts at the measuring jets of the gage plug. The rate of air flow is proportional to the clearance between the gage plug and test piece, which is indicated by the position of the float in the column.

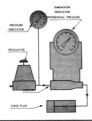

Venturi System

Filtered air passes through a regulator and is controlled at an optimum pressure. It then flows into a large Venturi chamber that converges into a smaller Venturi chamber and on through the measuring orifices. Each Venturi chamber is connected to opposite sides of a differential pressure sensitive device which drives an indicator. Clearance between the gage plug and test piece is proportional to air velocity and pressure changes in the two Venturi chambers, which causes the indicator to register.

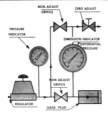

Differential System

Filtered air passes through a regulator and is controlled at an optimum pressure. It then passes into two separate channels, each containing a fixed master orifice. In one channel, the air is exhausted at the measuring jets of the gage plug. In the other, it is exhausted through an adjustable valve which can fix the reference indication—the meter "zero" —of a differential pressure sensitive device connected across the two channels as shown.

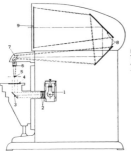

PICTORIAL REPRESENTATION OF THE PATH OF LIGHT THROUGH A CONTOUR MEASURING PROJECTOR

1. Light source
2. Condensing lenses
3. Mirror
4. Glass test piece support
5. Object
6. Projection lens
7. Prism
8. Mirrors
9. Screen

OPTICAL PROJECTION EQUIPMENT FACTS

Most optical projectors enjoy the following advantages over other direct-viewing magnification systems:

• The field of view is large, permitting more of the test piece to be seen at one setting.

• Two or more viewers can see the same image at the same time.

• Working space is usually much larger.

Projectors generally have greater table travel for making measurements over much greater areas.

The first commercial projectors for inspection purposes were developed during World War I. Because these instruments were applied for comparison gaging to fixed outlines on frosted screens, projectors became commonly known as optical comparators.

The components of an optical projector include:

• Lens System   • Illumination
• Screen   • Test Piece

In addition, most optical projectors include a collimator lens and devices for holding and positioning the test piece.

ILLUMINATION
It is desirable to obtain maximum possible light with as small a filament as possible.

COLLIMATOR LENS
This lens must pickup as much light as possible from the lamp and project it "straight and parallel" across the outline or surface to be inspected.

VIEWING LENS
This lens is designed to pickup the test piece shadow and surrounding illumination and pass them on to the viewing screen. Since the viewing screen is at a greater distance from the lens than the lens is from the test piece or test piece image, the screen image will appear magnified.

RELAY LENS
This lens relays the test piece image to an optimum position along the optical axis for subsequent magnification by the viewing lens. It is capable of projecting images "in air" at 1:1 magnification.

MIRRORS
With the use of mirrors, the lens system can be considerably more compact, making the image and test piece conveniently located for an operator.

MAGNIFICATION
The screen image is magnified in proportion to the focal length of the projection lens. Lenses having decimal magnifications of 31.25x and 62.5x were developed during earlier days when projectors were used chiefly as comparators. Every .001" on a test piece, when magnified by 31.25x, becomes 1/32" on the screen and can be measured with an ordinary steel rule. At 62.5x, each .001" is equal to 1/16" and can be scaled similarly.

SCREENS
Most screens are made of round, translucent glass, which represents a flat layout surface for outlining measurements or comparisons without interfering with the projected image.

TEST PIECE LOCATORS
A means for holding and accurately positioning the test piece is often provided to aid in bringing the desired contour into the field of view, and results in coordinated movement and measurement at the screen.

SURFACE PROJECTION
In addition to shadow projection, a projector is often equipped to present an enlarged surface view of a test piece. If sufficient light can be directed against the part, and if the surface is such that it will reflect an adequate amount of light, a "picture" view can be obtained.

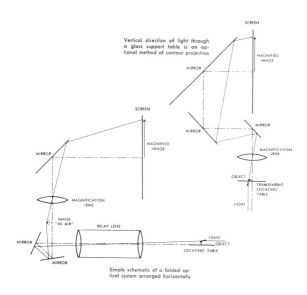

Vertical direction of light through a glass support table is an optional method of contour projection

SCREEN

MAGNIFIED IMAGE

MIRROR

SCREEN

MIRROR

MIRROR

MAGNIFIED IMAGE

MAGNIFICATION LENS

MIRROR

MAGNIFICATION LENS

OBJECT

TRANSPARENT LOCATING TABLE

LIGHT

IMAGE "IN AIR"

RELAY LENS

MIRROR

LIGHT

OBJECT

LOCATING TABLE

MIRROR

Simple schematic of a folded optical system arranged horizontally

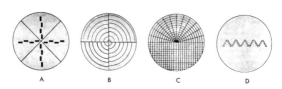

A   B   C   D

Projection screen charts are available with a number of standard and special outlines and configurations. Typical of the most common are (a) centerline references, (b) standard radii, (c) comparison grids and angles, and (d) screw thread forms.

COMMON METHODS and COMMON ERRORS
of ROUNDNESS MEASUREMENT

1. Diametric Measurement
—using size-measuring instruments.

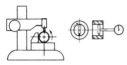

Limitations: This method detects only even-numbered lobing. Since odd-numbered lobing is very common, these measurements are unreliable; for constant-diameter parts with odd-numbered lobing are often badly out-of-round.

2. V-Block Measurement
—using size-measuring instruments with a V-angle in the anvil.

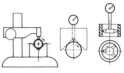

Limitations: With a V-block, ovality and other even-numbered lobing may be demagnified, or may not be indicated at all. For odd-numbered lobing, a different V-angle is needed for each frequency of lobing; and irregular lobing is difficult to assess.

Also, with ID's this method is not practical except with fixed-angle 3-point internal gages, which ignore many types of lobing.

3. Bench Center Radial Measurement
—using bench centers and an indicator.

Limitations: This method is not suitable for most ID's. For OD's, accuracy is affected by the shape, angle and alignment of centers and center holes, straightness of the part, and other factors—all of which must be controlled much more closely than the desired accuracy of measurement. Also, center and part errors may mask each other.

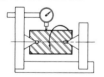

4. Radial Measurement with Ultra-Precision Spindle
—using a suitable indicator or recorder.

This equipment identifies and measures all types of out-of-roundness. With it, the only limitation is spindle accuracy; and roundness measurements accurate to 0.000,001" (one millionth of an inch) are provided

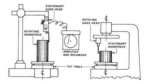

DIMENSIONING AND TOLERANCING DEFINITIONS

1. **Allowance** – An allowance is a prescribed difference between the maximum material condition of mating parts. It is the minimum clearance (positive allowance) or maximum interference (negative allowance) between such parts (see definition of fit).

2. **Clearance** – Clearance is the total space between mating parts.

3. **Dimension** – A dimension is a numerical value expressed in appropriate units of measure. It is indicated on drawings in conjunction with lines, symbols, and notes to define the geometrical characteristics of an object.

4. DIMENSIONING
   4.1 **Angular Dimensioning** – The angular dimensioning system is a method for indicating the position of a point, line, or surface by means of linear dimensions and angles, other than 90 degree angles.

   4.2 **Rectangular Dimensioning** – The rectangular dimensioning system is a method for indicating distances, locations, and sizes by means of linear dimensions measured parallel to reference lines or planes which are perpendicular to each other.

5. **Eccentricity** – Eccentricity is a condition where the axis of a particular feature is parallel to, but offset from, the axis of another feature; or where the axis of a rotating part mounted in an assembly does not coincide with the axis of the part about which it turns.

6. **Feature** – Features are specific characteristics or component portions of a part, which may include one or more surfaces such as holes, screw threads, profiles, or rabbets.

7. **Fit** – Fit is the general term used to signify the range of looseness or tightness which may result from the application of a specific combination of allowances or tolerances in the design of mating parts.

   7.1 **Actual Fit** – The actual fit between two or more mating parts is the relation existing between them with respect to the amount of clearance or interference which is present when they are assembled.

   7.2 **Clearance Fit** – A clearance fit is one having limits of size so prescribed that clearance always results when mating parts are assembled.

   7.3 **Interference Fit** – An interference fit is one having limits of size so prescribed that an interference always results when mating parts are assembled.

   7.4 **Transition Fit** – A transition fit is one having limits of size so prescribed that either a clearance or an interference may result when parts are assembled.

8. **Interchangeability** – Interchangeability is a condition of design wherein any and all mating parts will assemble and function properly without the need for any part modification at assembly.

OPTICS

The expanding use of optics to perform functions previously performed by mechanical or electrical methods has led to the design of new optical instruments and equipment. The following general information is furnished to provide guidance, direction, and technical assistance for daily use.

Units of dimensional measure may be in inches or milimeters. Some lens specs give radius of curvature and/or lens focal length in diopters. One diopter is the power of a lens whose focal length is one meter. There are numerous optical materials which are outlined in military specifications for most standard applications. (See typical specifications in other section of Handbook.)

Index of refraction of optical glasses is specified to the fourth decimal point rounded off to the third decimal place, with the following tolerances:

Below 1.600		$\pm$ .0010
From 1.600	To 1,730,	$\pm$ .0015
Above 1.730		$\pm$ .0020

The Abbe Constant is stated to the first decimal place and has the following tolerances:

Below 30		$\pm$ .2
From 30	To 45,	$\pm$ .3
From 45	To 55,	$\pm$ .4
Above 55		$\pm$ .5

There are four material grades:

GRADE A glass shall contain no visible striae, streaks or cords.

GRADE B glass shall only contain striae which are light and scattered when viewed in the direction of maximum visibility.

GRADE C glass shall only contain striae which are light when viewed in the direction of maximum visibility and are parallel to the face of the plate.

GRADE D glass contains more and heavier striae than contained in Grade C. The striae shall be parallel to the face of the plate.

Grades A and B are usually specified for components in precision viewing or projecting equipment, while Grades C and D may be used where less quality is required, such as in condenser optics. For very high quality windows and lenses, such as in Schliern optical systems, you can get a Grade AA glass on special order.

Defects - The various glass defects are:

STRIAE:
Streaks or layers of different indices of refraction within the glass, defined as striae, affect the sharp image definition of prisms and lenses

STRAIN:
Strain occurs in the material because of stresses incurred by improper annealing; but it can also be introduced during the grinding and polishing operations.

BUBBLES AND INCLUSIONS:
During manufacture, some small bubbles and inclusions occur and remain in the glass. As a maximum, they should not exceed 0.02% of the volume. Inclusions can also introduce stress and striae to the surrounding glass.

Quality - Table III gives a "rule-of-thumb" tolerance breakdown for various classes of lenses, prisms and flats. Use these values only as a guide.

TABLE III TOLERANCES FOR CLASSES
OF WORK AND TYPE OF PIECE

A. LENS

	Low Cost	Commercial	Precision	Extra Precise
Diameter	1. ±0.008	2. ±0.003	3. ±0.0008	4. ±0.0004
Centering	±1/2°	±6 mins.	±30 secs.	As req'd.
Thickness	±0.020	±0.010	±0.004	±0.002
Radius	To gage	±0.2%	±0.2%	As req'd. <±0.2%
Figure	To gage	3 Rings/ in. dia.	2 Rings/ in. dia.	1 Ring/ in. dia.
Surface Quality	120 - 80	80 - 50	60 - 40	As req'd.

B. PRISMS

Angles	±1/2°	±1/4°	±10 secs.	±2 secs.
Surface Quality	120 - 80	80 - 50	60 - 40	30 - 20
Figure	10λ/inch	2λ/inch	1/4λ/inch	As req'd.
Linear	±1/64	±0.005	±0.002	As req'd.

C. FLATS

Wedge	Factory Run Plate	0.001"/3"	10 secs.	As req'd.
Surface Quality	120 - 80	80 - 50	60 - 40	As req'd.
Figure	10λ/inch	2λ/inch	1/4λ/inch	As req'd.
Linear	±1/64	±0.005	±0.002	As req'd.

The surface quality specification, such as 80-50, refers to the scratch and dig standards of MIL-O-13830A. Table IV lists the quality code as defined by MIL-O-13830A, while Table V lists surface quality recommendations. The first number defines the maximum scratch width in 1/1000 mm (e.g., for 80, maximum width is 0.08 mm). Combined length of the heaviest scratches shouldn't exceed 1/4 the lens diameter. The second number defines the maximum diameter of digs and bubbles on the surface and within the glass in 1/100 mm (e.g., for 50, maximum diameter of a bubble is 0.5 mm). The permissible number of maximum size digs shall be one per each 20 mm of diameter or fraction thereof on any single optical surface. The sum of the diameters of all digs shall not exceed twice the diameter of the maximum size specified.

TABLE IV—SCRATCH AND DIG (BUBBLE) STANDARDS

(Based on MIL-O-13830A)

Scratch or Bubble No.	Scratch				Dig or Bubble			
	1		2		3		4	
	MM	Inch		MM	Inch	MM	Inch	
120	0.12	0.0047	1/4	1.20	0.0473	20	0.787	
80	0.08	0.0031	1/4	0.80	0.0315	20	0.787	
60	0.06	0.0024	1/4	0.60	0.0236	20	0.787	
40	0.04	0.0016	1/4	0.40	0.0158	20	0.787	
30	0.03	0.0012	1/4	0.30	0.0118	20	0.787	
20	0.02	0.0008	1/4	0.20	0.0079	20	0.787	
15	0.015	0.0006	1/4	0.15	0.0059	20	0.787	
10	0.010	0.0004	1/4	0.10	0.0039	1	0.040	
5	0.005	0.0002	1/4	0.05	0.0020	1	0.040	
3	0.003	0.00012	1/4	0.03	0.0012	1	0.040	

1. Maximum width of scratch.

2. A. (spherical): Portion of lens diameter that combined length of heaviest scratches do not exceed.

   B. (cylindrical): Portion of zone width (unless otherwise specified, zone is to be a square with one dimension equal to the width or width of clear area of lens) that combined length of heaviest scratches do not exceed.

3. Maximum bubble and dig diameter.

4. Distance that bubbles or digs must be separated in direction normal to line of sight.

NOTE: The permissible number of maximum size digs shall be one per each 20 mm (0.787") or diameter or fraction thereof on any single optical surface. The sum of the diameters of all digs as estimated by the inspector shall not exceed twice the diameter of the maximum size specified per 20 mm diameter. (Digs less than 2.5 microns shall be ignored.)

TABLE V. SURFACE QUALITY RECOMMENDATIONS

Item	Scratch	Dig or Bubble
Condensing Lens	80 to 120	60 to 80
Telescope Objective	60 to 80	40 to 50
Mirrors	30 to 80	30 to 50
Projection Lenses	40 to 60	20 to 40
Camera Lenses	40 to 60	20 to 40
Prisms	30 to 60	10 to 30
Microscope Objective	30 to 60	10 to 30
Oculars	15 to 30	5 to 20
Reticles	10 to 20	1 to 5

Unusual sizes or applications may require modified values. Table I of MIL-O-13830A lists values based on the beam diameter and location of the element.

Contrary to most non-optical measuring instruments, the
autocollimator operates without physical contact with
the work being measured. The autocollimator is a tele-
scope which provides collimated light and measures the
angle of its reflection from a plane mirror to a fraction
of an arc second - better than .000001 inches per inch.
It has found widespread use in precision engineering for
determination of straightness and squareness of machine
tools, as well as flatness of surface plates, in the precision calibration of rotary tables and dividing heads,
and in aircraft and missile technology to determine struc-
tural alignment and position or movement of guidance systems components.

OPTICAL MICROMETERS WITH SAMPLE READINGS

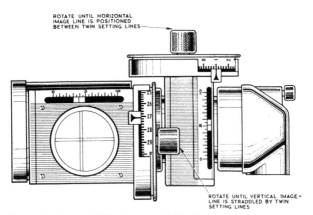

ROTATE UNTIL HORIZONTAL
IMAGE LINE IS POSITIONED
BETWEEN TWIN SETTING LINES

ROTATE UNTIL VERTICAL IMAGE-
LINE IS STRADDLED BY TWIN
SETTING LINES

Example: Horizontal displacement		Vertical displacement	
Micrometer =	27·1 sec.	Micrometer =	1 sec.
Turn counter =	3 min. 30 sec.	Turn counter =	5 min.
Reading	3 min. 57·1 sec.	Reading	5 min. 1 sec.

Relative Solderability of Metal Alloys and Coatings

Base Metal, Alloy or Applied Finish	Flux Requirements			Soldering Not Recommended
	Non-Corrosive	Corrosive	Special Flux and/or Solder	
Aluminum			X	
Aluminum-Bronze			X	
Beryllium				X
Beryllium Copper		X		
Brass	X	X		
Cadmium	X	X		
Cast Iron			X	
Chromium				X
Copper	X	X		
Copper-Chromium		X		
Copper-Nickel		X		
Copper-Silicon		X		
Gold	X			
Inconel			X	
Lead	X	X		
Magnesium			X	
Manganese-Bronze (High Tensile)				X
Monel		X		
Nickel		X		
Nichrome			X	
Palladium	X			
Platinum	X			
Rhodium		X		
Silver	X	X		
Stainless Steel			X	
Steel		X		
Tin	X	X		
Tin-Bronze	X	X		
Tin-Lead	X	X		
Tin-Nickel	X	X		
Tin-Zinc	X	X		
Titanium				X
Zinc		X		
Zinc Die Castings			X	

Solderability of Metals in Decreasing Order

1. Tin-Zinc Plate
2. Gold
3. Silver (Clean)
4. Cadmium Plate (Clean)
5. Copper (Clean)
6. Tin (Hot Dipped)
7. Tin Plate (Clean)
8. Solder Plate
9. Terne Plate
10. Lead
11. Cadmium Plate (Oxidized)
12. Copper (Oxidized)
13. Silver (Tarnished)
14. Tin Plate (Oxidized)
15. Nickel Plate
16. Brass
17. Bronze
18. Galvanized
19. Cast Iron
20. Steel (Mild)
21. Silicon Bronze
22. Alnico
23. Chromium
24. Inconel
25. Monel
26. Steel (Stainless)
27. Aluminum

When soldering dissimilar metals use one of the fluxes recommended for the metal which appears lower in the above list.

TEMPERATURE CONVERSION TABLE

In left column find known temperature in degrees C. or F.

Refer to corresponding C. or F. column for the equivalent.

	°Cent.	°Fahr.		°Cent.	°Fahr.		°Cent.	°Fahr.
−100	−73.3	−148	41	5.00	105.8	92	33.3	197.6
−90	−67.8	−130	42	5.56	107.6	93	33.9	199.4
−80	−62.2	−112	43	6.11	109.4	94	34.4	201.2
−70	−56.7	−94	44	6.67	111.2	95	35.0	203.0
−60	−51.1	−76	45	7.22	113.0	96	35.6	204.8
−50	−45.6	−58	46	7.78	114.8	97	36.1	206.6
−40	−40.0	−40	47	8.33	116.6	98	36.7	208.4
−30	−34.4	−22	48	8.89	118.4	99	37.2	210.2
−20	−28.9	−4	49	9.44	120.2	100	37.8	212.0
−10	−23.3	14	50	10.0	122.0	105	40.6	221.0
0	−17.8	32	51	10.6	123.8	110	43	230
1	−17.2	33.8	52	11.1	125.6	120	49	248
2	−16.7	35.6	53	11.7	127.4	130	54	266
3	−16.1	37.4	54	12.2	129.2	140	60	284
4	−15.6	39.2	55	12.8	131.0	150	66	302
5	−15.0	41.0	56	13.3	132.8	160	71	320
6	−14.4	42.8	57	13.9	134.6	170	77	338
7	−13.9	44.6	58	14.4	136.4	180	82	356
8	−13.3	46.4	59	15.0	138.2	190	88	374
9	−12.8	48.2	60	15.6	140.0	200	93	392
10	−12.2	50.0	61	16.1	141.8	210	99	410
11	−11.7	51.8	62	16.7	143.6	212	100	413
12	−11.1	53.6	63	17.2	145.4	220	104	428
13	−10.6	55.4	64	17.8	147.2	230	110	446
14	−10.0	57.2	65	18.3	149.0	240	116	464
15	−9.44	59.0	66	18.9	150.8	250	121	482
16	−8.89	60.8	67	19.4	152.6	260	127	500
17	−8.33	62.6	68	20.0	154.4	270	132	518
18	−7.78	64.4	69	20.6	156.2	280	138	536
19	−7.22	66.2	70	21.1	158.0	290	143	554
20	−6.67	68.0	71	21.7	159.8	300	149	572
21	−6.11	69.8	72	22.2	161.6	310	154	590
22	−5.56	71.6	73	22.8	163.4	320	160	608
23	−5.00	73.4	74	23.3	165.2	330	166	626
24	−4.44	75.2	75	23.9	167.0	340	171	644
25	−3.89	77.0	76	24.4	168.8	350	177	662
26	−3.33	78.8	77	25.0	170.6	360	182	680
27	−2.78	80.6	78	25.6	172.4	370	188	698
28	−2.22	82.4	79	26.1	174.2	380	193	716
29	−1.67	84.2	80	26.7	176.0	390	199	734
30	−1.11	86.0	81	27.2	177.8	400	204	752
31	−0.56	87.8	82	27.8	179.6	410	210	770
32	0	89.6	83	28.3	181.4	420	216	788
33	0.56	91.4	84	28.9	183.2	430	221	806
34	1.11	93.2	85	29.4	185.0	440	227	824
35	1.67	95.0	86	30.0	186.8	450	232	842
36	2.22	96.8	87	30.6	188.6	460	238	860
37	2.78	98.6	88	31.1	190.4	470	243	878
38	3.33	100.4	89	31.7	192.2	480	249	896
39	3.89	102.2	90	32.2	194.0	490	254	914
40	4.44	104.0	91	32.8	195.8	500	260	932

Thermal Expansion Coefficient

Material	Inch Per Inch Per Degree Fahrenheit
Aluminum	.000012
Brass	.000010
Bronze	.000010
Concrete	.000008
Copper	.000009
Cast Iron	.0000056
Lead	.000016
Magnesium	.000014
Nickel	.000007
Steel, Carbon	.0000063
Tungsten Carbide	.0000033
Granite	.0000044
Stainless Steel, Magnetic	.0000061
Stainless Steel, Non-Magnetic	.0000096
Glass, Plate	.000005
Glass, Pyrex	.0000018
Quartz Glass, Fused	.0000003
Invar	.0000009
Titanium	.000005
Plastics	.0002 Approx.

Conversion-Linear Measurement to Angles

Linear Measure	Per Inch	Per 10 Inches	Per Foot
.000 001"	0.206 Seconds	0.021 Seconds	0.017 Seconds
.000 025"	5.157 Seconds	0.516 Seconds	0.430 Seconds
.000 05"	10.3 Seconds	1.03 Seconds	0.86 Seconds
.000 1"	20.6 Seconds	2.06 Seconds	1.72 Second
.001"	3 Min. 26 Sec.	20.6 Seconds	17.2 Seconds
.005"	17 Min. 11 Sec.	1 Min. 43 Sec.	1 Min. 26 Sec.

(Courtesy of Scherr-Tumico, Inc.)

Conversion-Angles To Linear Measurement

Angles	Per Inch	Per 10 Inches	Per Foot
1 Second	.000 005"	.000 048"	.000 058"
5 Seconds	.000 024"	.000 242"	.000 291"
10 Seconds	.000 048"	.000 485"	.000 582"
20 Seconds	.000 097"	.000 970"	.001 163"
30 Seconds	.000 145"	.001 454"	.001 745"
1 Minute	.000 291"	.002 909"	.003 491"

Wire Gage Standards

─────────Decimal Parts of an Inch─────────

Wire Gage No.	American or B&S Gage	Birmingham or Stubs Wire Gage	Washburn Moen Steel Wire Gage	A.S.&W. Music Wire Gage	British Imp. Wire Gage	U. S. Std. Revised	U.S. Std. for Plate Gage
0000000	....	......	.4900	......	.500	......	....
000000	.580049	......	.4615	.004	.464	......	.46875
00000	.516549	.500	.4305	.005	.432	......	.4375
0000	.460	.454	.3938	.006	.400	......	.40625
000	.40964	.425	.3625	.007	.372	......	.375
00	.3648	.380	.3310	.008	.348	......	.34375
0	.3249	.340	.3065	.009	.324	......	.3125
1	.2893	.300	.2830	.010	.300	......	.28125
2	.25763	.284	.2625	.011	.276	....	.265625
3	.22942	.259	.2437	.012	.252	.2391	.250
4	.20431	.238	.2253	.013	.232	.2242	.234375
5	.18194	.220	.2070	.014	.212	.2092	.21875
6	.16202	.203	.1920	.016	.192	.1943	.203125
7	.1443	.180	.1770	.018	.176	.1793	.1875
8	.1285	.165	.1620	.020	.160	.1644	.171875
9	.11443	.148	.1483	.022	.144	.1495	.15625
10	.1019	.134	.1350	.024	.128	.1345	.140625
11	.090742	.120	.1205	.026	.116	.1196	.125
12	.080808	.109	.1055	.029	.104	.1046	.109375
13	.0720	.095	.0915	.031	.092	.0897	.09375
14	.0641	.083	.0800	.033	.080	.0747	.078125
15	.0571	.072	.0720	.035	.072	.0673	.0703125
16	.0508	.065	.0625	.037	.064	.0598	.0625
17	.0453	.058	.0540	.039	.056	.0538	.05625
18	.0403	.049	.0475	.041	.048	.0478	.050
19	.0359	.042	.0410	.043	.040	.0418	.04375
20	.0320	.035	.0348	.045	.036	.0359	.0375
21	.0285	.032	.0317	.047	.032	.0329	.034375
22	.0253	.028	.0286	.049	.028	.0299	.03125
23	.0226	.025	.0258	.051	.024	.0269	.028125
24	.0201	.022	.0230	.055	.022	.0239	.025
25	.0179	.020	.0204	.059	.020	.0209	.021875
26	.0159	.018	.0181	.063	.018	.0179	.01875
27	.0142	.016	.0173	.067	.0164	.0164	.0171875
28	.0126	.014	.0162	.071	.0148	.0149	.015625
29	.0113	.013	.0150	.075	.0136	.0135	.0140625
30	.0100	.012	.0140	.080	.0124	.0120	.0125
31	.0089	.010	.0132	.085	.0116	.0105	.0109375
32	.0080	.009	.0128	.090	.0108	.0097	.01015625
33	.0071	.008	.0118	.095	.0100	.0090	.009375
34	.0063	.007	.0104	.100	.0092	.0082	.00859375
35	.0056	.005	.0095	.106	.0084	.0075	.0078125

SYMBOLS PER ANSI STANDARD
Y 14.5M 1982

CHARACTERISTIC	ANSI-Y14.5
STRAIGHTNESS	⏤
FLATNESS	⏥
ANGULARITY	∠
PERPENDICULARITY (SQUARENESS)	⊥
PARALLELISM	//
CONCENTRICITY	◎
POSITION	⌖
CIRCULARITY (ROUNONESS)	○
SYMMETRY	USE ⌖
PROFILE OF A LINE	⌒
PROFILE OF A SURFACE	⌓
RUNOUT (CIRCULAR)	↗ ↗
RUNOUT (TOTAL)	⌰ ⌰
CYLINDRICITY	⌭
DATUM FEATURE	-A-
MAXIMUM MATERIAL CONDITION (MMC)	Ⓜ
REGARDLESS OF FEATURE SIZE (RFS)	Ⓢ
LEAST MATERIAL CONDITION (LMC)	Ⓛ
DATUM TARGET	A1
PROJECTED TOLERANCE ZONE	Ⓟ

Updated 7/84

TOLERANCE RANGE OF MACHINING PROCESSES

- Lapping & Honing
- Grinding, Diamond Turning & Boring
- Broaching
- Reaming
- Turning, Boring, Slotting, Planing, & Shaping
- Milling
- Drilling

TOLERANCES

RANGE OF SIZES From	To & incl.									
.000	.599	.00015	.0002	.0003	.0005	.0008	.0012	.002	.003	.005
.600	.999	.00015	.00025	.0004	.0006	.001	.0015	.0025	.004	.006
1.000	1.499	.0002	.0003	.0005	.0008	.0012	.002	.003	.005	.008
1.500	2.799	.00025	.0004	.0006	.001	.0015	.0025	.004	.006	.010
2.800	4.499	.0003	.0005	.0008	.0012	.002	.003	.005	.008	.012
4.500	7.799	.0004	.0006	.001	.0015	.0025	.004	.006	.010	.015
7.800	13.599	.0005	.0008	.0012	.002	.003	.005	.008	.012	.020
13.600	20.999	.0006	.001	.0015	.0025	.004	.006	.010	.015	.025

Plain Cylindrical Plug and Ring Gage Tolerances
American Gage Design Standard

Nominal Size (inches)		Gagemakers' Tolerance Classes (microinches)					
above	to and including	XXX	XX	X	Y	Z	ZZ
.029	.825	10	20	40	70	100	200
.825	1.510	15	30	60	90	120	240
1.510	2.510	20	40	80	120	160	320
2.510	4.510	25	50	100	150	200	400
4.510	6.510	32.5	65	130	190	250	500
6.510	9.010	40	80	160	240	320	640
9.010	12.010	50	100	200	300	400	800

VIBRATION TIPS

SINUSOIDAL

$$DA = \frac{g}{0.0511f^2}$$

A = 0.0511f²DA

DA - Double Amplitude

A - Acceleration

f - Frequency

.010 Per Div.  .004 Per Div.

.020 Per Div.

TOTAL DISPLACEMENT

With the use of a wedge, the accelerometer may be checked. Listed
below are two typical points.

56 HERTZ @ .060 DA = 10 g's
44 HERTZ @ .100 DA = 10 g's

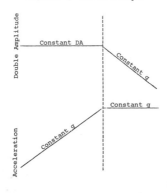

UNITS OF MEASURE

Metric System Prefixes

Mega = 1,000,000   Deci = 0.1
Kilo = 1,000   Centi = 0.01
Hecto = 100   Milli = 0.001
Deka = 10   Micro = 0.000001

Length

1 centimeter	= 0.3937 inches	= 0.0328 feet
1 meter	= 39.37 inches	= 1.0936 yards
1 kilometer	= 0.62137 miles	= 3280 feet
1 inch	= 2.54 centimeters	
1 foot	= 0.3048 meters	
1 mil	= 0.001 inch	

Square Measure

1 sq. cm.	= 0.1550 sq. in	
1 sq. meter	= 1.196 sq. yds.	= 10.784 sq. ft.
1 sq. kilometer	= 0.386 sq. miles	
1 sq. inch	= 6.452 sq. centimeters	
1 sq. foot	= 929.03 sq. cm.	= 0.092903 sq. meters
1 sq. yard	= 0.8361 sq. meters	
1 sq. mile	= 2.59 sq. kilometers	
1 circular mil	= 0.7854 sq. mils	
1 sq. inch	= 1,000,000 sq. mils	

Cubic Measure

1 cu. centimeter	= 0.061 cu. inch 1 cu. in. = 16.39 cu. cm.
1 cu. meter	= 1.308 cu. yds = 35.316 cu. feet
1 gallon (U.S.)	= 231 cubic inches
1 cu. ft.	= 7.48 gallons 1 liter = 1000 cu. centimeters

Time

1 day = 86,400 seconds   1 year = 8760 hours (approx.)

Mass

1 slug	= 32.2 pounds mass = 14.606 kilograms
1 pound mass	= 453.6 grams

Force

1 pound force	= 1 slug x 1 foot sec/sec
1 dyne	= 1 gram x 1 centimeter/sec/sec
1 newton	= 1 kilogram x 1 meter/sec/sec
1 pound force	= 4.452 newtons
1 newton	= 100,000 dynes = 0.224 pounds force
1 gram force	= 980.6 dynes

Pressure

1 atmosphere = 14.69 pounds/sq. inch = 29.92 in. of Hg.
      = 76 cm of Hg = 33.9 ft. of water
1 in. Hg. = 0.491 pounds/sq. inch
Water pressure pounds/sq. inch = head in ft. x 0.434

Work and Energy—Mechanical

1 erg = 1 dyne   x 1 centimeter
1 joule = 1 newton x 1 meter = 10^5 dynes x 10^2 cm = 10^7 ergs
1 ft lb = 1 pound force x 1 foot = 1.356 joules

Work and Energy—Heat Equivalent

1 Btu raises 1 pound of water 1° F
1 gram calorie raises 1 gram of water 1°C
1 Btu = 252 gram calories = 778 ft lb = 1055 joules
1 gram calorie = 0.003964 Btu = 4.184 joules
1 horsepower hour = 2544 Btu

Work and Energy—Electrical Equivalent

1 joule = 1 watt x 1 second = 1 amp (dc) x 1 volt (dc) x 1 sec
W (joules) = ½ L (henries) x I (amperes)²
W (joules) = ½ C (farads) x E (volts)²
1 kilowatt hour = 3,600,000 joules

Power

1 watt = 1 joule/sec
1 horsepower = 550 ft lb/sec = 746 watts
1 watt = 3.412 Btu/hr = 0.239 gram calorie/sec
P (watts) = R (ohms) x I (amperes)²
P (watts) = $\dfrac{E \text{ (volts)}^2}{R \text{ (ohms)}}$

Angles

1 circle = 2π radians = 360 degrees
1 radian = 57.3 degrees   1 degree = 0.01745 radians

Geometric Figures

Circle, area of = D^2 x 0.7854 = πr^2 r = radius
Circle, circumference of = πD or 2πr
Sphere, area of = πD = 4πr^2 D = diameter
Sphere, volume of = D^3 x 0.5236 = 4/3πr^3
Triangle, area of = ½ altitude x base
Cone, volume of = area of base x ⅓ altitude
Trapezoid, area of = ½ (sum of parallel sides) x altitude
Pyramid, volume of = area of base x ⅓ altitude

Miscellaneous Constants

π = 3.14159   e = 2.71828

$\log_e X$ = 2.30259 $\log_{10} X$

Electronic charge = 4.8 x 10^{-10} e.s.u. = 1.60 x 10^{-20} e.m.u.
Mass units = 1.07 x 10^{-3} x Mev = 6.71 x 10^4 ergs
Speed of light = 3 x 10^8 meters/sec
Speed of sound (in air at sea level) = 1100 ft/sec

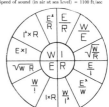

76

DECIMAL EQUIVALENTS and TAP DRILL SIZES

of Wire Gauge, Letter and Fractional Size Drills

(Tap Drill Sizes Based on 75% Maximum Thread)

Fract. Size Drills Inches	Wire Gauge & Letter Size Drills	Decimal Equiv. Inches	Size of Thread	Threads Per Inch
	80	.0135		
	79	.0145		
1/64		.0156		
	78	.0160		
	77	.0180		
	76	.0200		
	75	.0210		
	74	.0225		
	73	.0240		
	72	.0250		
	71	.0260		
	70	.0280		
	69	.0292		
	68	.0310		
1/32		.0312		
	67	.0320		
	66	.0330		
	65	.0350		
	64	.0360		
	63	.0370		
	62	.0380		
	61	.0390		
	60	.0400		
	59	.0410		
	58	.0420		
	57	.0430		
	56	.0465		
3/64		.0469	0	80
	55	.0520		
	54	.0550		
	53	.0595	1	64 / 72
1/16		.0625		
	52	.0635		
	51	.0670		
	50	.0700	2	56 / 64
	49	.0730		
	48	.0760		
5/64		.0781		
	47	.0785	3	48
	46	.0810		
	45	.0820	3	56
	44	.0860		
	43	.0890	4	40
	42	.0935	4	48
3/32		.0937		
	41	.0960		
	40	.0980		
	39	.0995		
	38	.1015	5	40
	37	.1040	5	44
	36	.1065	6	32
7/64		.1094		
	35	.1100		
	34	.1110		
	33	.1130	6	40
	32	.1160		
	31	.1200		
1/8		.1250		
	30	.1285		
	29	.1360	8	32 / 36
	28	.1405		
9/64		.1406		
	27	.1440		
	26	.1470		
	25	.1495	10	24
	24	.1520		
	23	.1540		
5/32		.1562		
	22	.1570		
	21	.1590	10	32
	20	.1610		
	19	.1660		
	18	.1695		
11/64		.1719		
	17	.1730		
	16	.1770	12	24
	15	.1800		
	14	.1820	12	28
	13	.1850		
3/16		.1875		
	12	.1890		
	11	.1910		
	10	.1935		
	9	.1960		
	8	.1990		
	7	.2010	1/4	20
13/64		.2031		
	6	.2040		
	5	.2055		
	4	.2090		
	3	.2130	1/4	28
7/32		.2187		
	2	.2210		
	1	.2280		
	A	.2340		
15/64		.2344		
	B	.2380		
	C	.2420		
	D	.2460		
1/4		.2500		
	E	.2500		
	F	.2570	5/16	18
	G	.2610		
17/64		.2656		
	H	.2660		
	I	.2720	5/16	24
	J	.2770		
	K	.2810		
9/32		.2812		
	L	.2900		
	M	.2950		
19/64		.2969		
	N	.3020		
5/16		.3125	3/8	16
	O	.3160		
	P	.3230		
21/64		.3281		
	Q	.3320	3/8	24
	R	.3390		
11/32		.3438		
	S	.3480		
	T	.3580		
23/64		.3594		
	U	.3680	7/16	14
3/8		.3750		
	V	.3770		
	W	.3860		
25/64		* .3906	7/16	20
	X	.3970		
	Y	.4040		
13/32		.4062		
	Z	.4130		
27/64		.4219	1/2	13
7/16		.4375		
29/64		.4531	1/2	20
15/32		.4687		
31/64		.4844	9/16	12
1/2		.5000		
33/64		.5156	9/16	18
17/32		.5312	5/8	11
35/64		.5469		
9/16		.5625		
37/64		.5781	5/8	18
19/32		.5937		
39/64		.6094		
5/8		.6250		
41/64		.6406		
21/32		.6562	3/4	10
43/64		.6719		
11/16		.6875	3/4	16
45/64		.7031		
23/32		.7187		
47/64		.7344		
3/4		.7500		
49/64		.7656	7/8	9
25/32		.7812		
51/64		.7969		
13/16		.8125	7/8	14
53/64		.8281		
27/32		.8437		
55/64		.8594		
7/8		.8750	1	8
57/64		.8906		
29/32		.9062		
59/64		.9219		
15/16		.9375	1	14
61/64		.9531		
31/32		.9687		
63/64		.9844	1 1/8	7
1		1.0000		

TYPICAL PROCESS CONTROL CHART

A typical process control chart you can use to plot any given dimension with stated tolerances. Measurements should be taken on schedule by Quality Control Inspectors.

ABC Specification: 1.850 + or − .010

1.861	Out of Control
1.860	
	Upper Control Limit "Caution Zone"
1.855	
1.854	Notify Production Operator of Trend
	"Safety Zone"
1.850	Set Up
	Set Up
	"Safety Zone"
1.846	Notify Production Operator of Trend
1.845	
	Lower Control Limit
	"Caution Zone"
1.840	
1.839	

Out of Control

Example: Always set up the machine or process as close to actual 1.850 in order to utilize all of the given tolerance and stay within the upper and lower control safety limits. Report trends immediately to the production operator and to supervision. Always Stay Within the 1.840 and 1.860 for Actual Control. Remember, if you set up the machine or process at the upper or lower control limit you will be in trouble before you start because you have used your tolerances one way or the other against yourself and not to your benefit. Each Dot On The Chart Defines Actual Measurement Made Every 15 Minutes By The Quality Control Inspector.

JOB INSTRUCTION TRAINING

WHEN TO INSTRUCT ...

1. When developing a skill

2. When imparting information

3. When developing an attitude

GET READY TO INSTRUCT ...

1. Know the operations yourself

    A. Quality, quantity, method, techniques, etc.

2. Break down the planned operation

    A. List movements in proper order for clarity

    B. Stress the key points & safety precautions

3. Get everything ready

    A. Plan of instruction

    B. Tools, work equipment, materials, supplies, etc.

4. Arrange the work station/work area

    A. As the worker is expected to keep it!

5. Approach the learner

    A. Friendly, helpful, patient attitude

    B. He or she wants to learn!

JOB INSTRUCTION TRAINING

HOW TO INSTRUCT ...

1. Preparation of the learner

   A. Put learner at ease, break the ice, be friendly

   B. State the job & find out what he/she knows

   C. Explain the purpose and importance & profit

   D. Place the learner in proper position to observe your demonstrations

2. Presentation to learner of skill & information

   A. Show, tell, illustrate & question

   B. Explain ONE point at a time, slowly & patiently

   C. Stress key factors & safety

   D. Check, question, repeat part or ALL as needed

   E. Be sure the learner really learns

3. Application and performance by the learner

   A. Have learner apply the skill or information

   B. Correct errors, & encourage the learner

   C. Ask questions, starting with Why, Who, When, Where, What, How

   D. Observe performance, repeat instruction if needed

   E. Continue until YOU know that HE/SHE knows!

4. Follow-up and supervise

   A. Put the learner on his/her own

   B. Have the learner perform skill or apply information

   C. Check, encourage and invite questions

   D. Taper off close supervision as learner progresses

   E. Audit the learner from time to time and don't forget to express your personal interest in his/her progress. **Praise good work.**

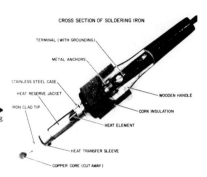

CROSS SECTION OF SOLDERING IRON

TERMINAL (WITH GROUNDING)

METAL ANCHORS

STAINLESS STEEL CASE

HEAT RESERVE JACKET

IRON CLAD TIP

WOODEN HANDLE

CORK INSULATION

HEAT ELEMENT

HEAT TRANSFER SLEEVE

COPPER CORE (CUT AWAY)

OPTIONAL: ➡
Dielex Coating
for Servicing
Hot Equipment

Controlled Heat Soldering Station

UNCONTROLLED SOLDERING IRONS CAN CAUSE COMPONENT FAILURES

Thus, the Controlled Heat Soldering Iron & Station is Recommended.

1. 3 Wire Cords for safety.
2. Positive Tip Ground.
3. No Magnetic Fields created
4. Leakage Rate of no more than 2 Milivolts
5. No more than 10 degrees F. above controlled temperature in idle.

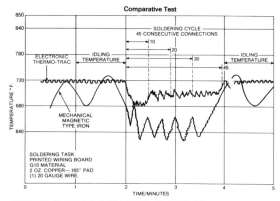

Comparative Test

SOLDERING CYCLE
45 CONSECUTIVE CONNECTIONS

ELECTRONIC THERMO-TRAC

IDLING TEMPERATURE

IDLING TEMPERATURE

MECHANICAL MAGNETIC TYPE IRON

SOLDERING TASK
PRINTED WIRING BOARD
G10 MATERIAL
2 OZ. COPPER—.165" PAD
(1) 20 GAUGE WIRE.

TEMPERATURE °F.

TIME/MINUTES

THERM-O-TRAC SOLDERING STATIONS offer the ultimate in precision, performance, and protection:

- Optimum temperature can be easily dialed, 500 to 850° F—no hot tips to change.
- Minimum temperature drop from idling to load conditions.
- Maximum temperature stability during work cycle.
- No magnetic field—safe for sensitive circuitry.
- Neutral switching eliminates danger of RF interference or voltage spikes generated by electro-mechanical switching.
- Protection to sensitive components thru isolated low voltage supply and posi-ground zero voltage system.
- All parts replaceable—can be repaired in your plant.
- No moving parts to fail, anneal, or stick—all solid state.
- Iron separate from controls—less danger of damage to delicate circuits from dropping or striking work bench.
- Variety of standard tips available in popular sizes and shapes.
- Heavy iron plating and precise temperature control insures long tip life.

STEDI-HEAT SOLDERING STATION

FOR LIGHT WORK
    Dependable performance
    Safe from overheating

COMMON APPLICATIONS:

Touch-Up	Diodes
P/C Board	Edge Connectors
Integrated Circuits	

SPECIFICATIONS:

Idle Temperature 650°F	Case Length 2-1/4"
Tip Diameter 3/16"	Case Diameter 3/8"
Tip Reach 1"	Weight 1-1/4 oz.
(Using Standard 2-1/4" Tips)	Hi Temp—Hi Flex Cord
O.A.L. 7" (Less Tip)	

TYPICAL SOLDERING TASK

P/C Board 1/16"	Continuous soldering
Pad .165" Dia.—2 oz. Copper	Rate .04 min. per connection
With crimped 20 ga. Copper Lead	Tip HT207D

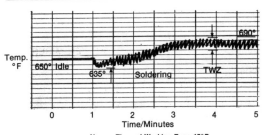

Narrow Thermal Working Zone 15°F
No overheating in stand

PROTECTING AND HANDLING COMPONENTS AND ASSEMBLIES
SUBJECT TO STATIC DAMAGE

The following precautions must be taken to keep valuable components
and assemblies from being damaged by Static Electric Discharge;
Poorly Grounded and Uncontrolled Soldering Irons; Work Stations
and/or People.

Receiving and Quality Assurance Receiving Inspection

Parts will be delivered in many different ways. Some methods of
shipping will be as follows:

1. Components mounted on black coated styrofoam.
2. Components mounted on black sponge.
3. Components contained in special I.C. Tubes
4. Components packaged individually on black sponge within a
    plastic container.

The parts may or may not be identified as being subject to Static
damage. Therefore, Receiving, Q.A. Inspection and stock handlers
must become familiar with parts requiring special handling. Training, guidance, direction and attitudes will play a major role in
the success of this Total Prevention Control and Assurance Program.

Storage of Components

1. Storage bins should be marked to indicate parts are to receive
    special handling because of created Static Damage.
2. Parts should be kept in part number order or segregated into a
    special group. They must be identified as requiring special
    handling because of damage that can be caused by "Handling"

Stock Handlers

1. Parts must remain on anti-static material at all times.
2. Parts must be picked up by the body of the component without
    touching the leads, if being transferred to anti-static mate-
    rial in another container.
3. Parts must never be placed in plastic bags or put in boxes
    without anti-static protection.

Assembly: The individual and the Work Station should be grounded
           via Ground Straps.

1. Parts on anti-static material must never be removed until
    inserted into a printed wiring board assembly.
2. An I.C. Component inserter should be used to pick up the I.C.
    and control the leads for insertion into a P.W.B.

Assembly (Cont.)

3. A Posi-grounded soldering iron must be used to solder components. The Hexacon Therm-O-Trac Soldering Stations and Irons are a primary source.
4. During assembly the P.W.B. should be on anti-static material and the individual should be grounded.
5. During the cleaning operation the P.W.B. assembly should be attached to a grounded strap while in the degreaser/cleaner.
6. The P.W.B. assembly should be placed in a anti-static bag.

Test Technicians: The individual should be grounded via Ground Straps.

1. The P.W.B. should be received in a anti-static bag. When removing the P.W.B. for test, ground yourself. After test return P.W.B. to anti-static bag.
2. If P.W.B. is to be repaired, soldering must be done with a Posi-grounded soldering iron. See Assembly 3.

Storage of Assemblies

1. P.W.B. is to be stocked in an a anti-static bag or material. No plastic bag or bubble pack is permitted.

Packing and Shipping

1. P.W.B. should be presented for shipment in anti-static bag or material.
2. P.W.B. is to be packed in insta-pack. Plastic peanuts/bubble pack, etc. are not permitted.
3. Assemblies containing PROM's or ROM's need to be wrapped in aluminum foil to protect against X-Rays.

Field Service Technicians

Great care must continue to be taken to insure a reliable, quality working P.W.B. in the field. You can ruin components and the whole assembly while trying to install the printed wiring board. Please be careful by:

1. Never work on carpet. You generate static electricity.
2. Ground yourself before removing P.W.B. from a system or from anti-static bag/material/package, etc.
3. Immediately install P.W.B. into system and place P.W.B. as removed from the system, into anti-static bag and return it for repair, testing, etc. per above instructions.

CONDUCTO-FOAM I — Soft conductive foam, ideal for storing and shipping parts requiring both static electrical and mechanical protection. Foam equalizes static potential around parts and assemblies. Available in flat, convoluted and special shapes. Base foam meets MIL-P-26514 Type I, Class II. Order W-100.

WESTRAPS — Personal grounding straps for assembly workers, maintenance technicians, medical personnel, etc. Westraps allow free movement while eliminating the body capacitance effects involved in static transmission. High resistance to ground protects wearer against AC potentials. Order 0-0741.

CONDUCTO-FOAM II — Firm conductive foam for tabletops, shipping or storage. Makes workbenches safe for handling easily shorted devices. For safe transport of MOS and other pinned devices, simply press leads into piece of conductive foam. Available in many shapes. Foam meets MIL-P-26514. Order W-1100.

GROUNDING STRAPS — Conductive plastic safely drains static electricity to ground from table top sheeting, seat covers, etc. High resistance protects personnel from line voltages while allowing static to bleed off to ground. Straps are up to 50 feet long, with clips at each end. Order W-0745.

DIP SHIPPING TUBES — Assure safe arrival of MOS devices and facilitate automatic testing and handling. Conductive materials equalize pin-to-pin potentials and prevent blowout. Available with slotted tops for easy counting of devices. Choose coated plastic, reusable Westat or aluminum. Order W-0600.

WESTAT SEAT COVERS — Snug-fitting anti-static covers for workstools and chairs prevent body motion and friction from generating high potential static charges. For optimum protection of sensitive devices, use with Grounding Straps, Westraps and Westat Sheeting. Order W-8000.

INTEGRATED CIRCUIT (DIP) COMPONENT SIDE OF BOARD

ACCEPTABLE

1. GOOD FILLET AROUND TERMINAL LEAD.

2. SOLDER IS BRIGHT AND SHINY.

MINIMUM ACCEPTABLE

1. MAXIMUM SOLDER USED.

2. MINOR SOLDER SPILLAGE ON TERMINAL LEAD.

3. GOOD WETTING ON ALL CONNECTIONS.

REJECT

1. EXCESSIVE SOLDER USED.

2. EVIDENCE OF CONTAMINATION IN SOLDER.

3. SOLDER NOT WETTED TO LEAD.

4. SOLDER PEAK ON LEAD.

INTEGRATED CIRCUIT (DIP) PIN SIDE OF BOARD

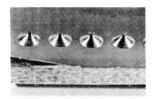

<u>ACCEPTABLE</u>

1. GOOD WETTING ON ALL LEADS

2. LEADS ARE EVENLY CUT

3. SOLDER IS SMOOTH, BRIGHT AND SHINY.

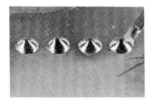

<u>MINIMUM ACCEPTABLE</u>

1. GOOD WETTING ON ALL LEADS.

2. SOLDER IS MAXIMUM, BUT LEADS STILL DEFINED.

3. SOLDER EXTENDS BEYOND ONE PAD, FOURTH LEAD FROM LEFT.

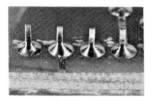

<u>REJECT</u>

1. PIN HOLE IN SOLDER.

2. INSUFFICIENT SOLDER USED ON ONE TERMINAL.

3. BASE METAL SHOWING ON LEAD.

4. LEADS CUT TOO LONG.

5. LEADS NOT PROPERLY CLEANED.

6. SOLDER SPATTER.

FINAL QUALITY ASSURANCE SYSTEMS TEST

AND INSPECTION RECORD

Procedure: The Q. A. Dept. may witness tests or conduct all or any tests
necessary to assure confidence in the System. The System Log
will reflect the System as documented and turned over to the
Field Service Installation Technician/Engineer. All discrepancies
found during the final Q. A. Test, Inspection or Audits conducted
as related to the System, shall be recorded below, corrected by
the assigned technician, and re-inspected for acceptance by Q. A.
All items must be completed, signed and dated, prior to final
acceptance by Q. A. The purpose and intent of this procedure
is to assure quality, confidence and corrective action. The
General Visual-Mechanical Insp./Test Checklist will be used
as a guide, in addition to any specific documentation related
to the System or Product.

Customer: _____ System: _____

Tested & Released to Q. A. by: _____ Date: _____

No.	Please Record Every Discrepancy -Problem-(Please be specific)	Noted By	Corrected By	Q.A. By

Tests, Inspections, Configuration Check, Documentation, All Records, in order
& signed off.

Q. A. REP._____ DATE_____

FIELD SERVICE REP._____ DATE_____

Dr. Carter

MURPHY'S LAW and OTHER THOUGHTS

1. Whatever strikes a fan is not evenly distributed....

2. Interchangeable parts, wont......

3. No good deed goes unpunished......

4. A $300.00 color picture tube will protect a 10¢ fuse
    by blowing first.......

5. Leakproof Seals, will........

6. Of two possible events, only the undesired one will
    occur........

7. Any wire, pipe or board cut to exact length will be
    1/4 inch too short.........

8. All warrantee's expire upon payment of the invoice.....

9. Nature always sides with the hidden flaw......

10. The professional Q.A. Auditor will always find the
    one reject in the entire lot.......

I am sure you can add to this brief list of 'Standards'...

11.

12.

13.

14.

15.

16.

17.

Dr. Carter's "FIVE PHASE QUALITY PROGRAM"

PHASE ONE: PREVENTION: The goal is to have a 'Formalized Prevention Program that will allow you to 'Prevent Quality & Related Problems'. You must have a documented &
functioning Quality Assurance Program/System to permit
you to discover problems & take corrective action before your products and or services enter the engineering, purchasing, manufacturing and field/customer service cycles. Prevention must start in the initial stage
of design & planning. This is how you prevent problems
at the lowest total cost, you cut them off at the pass,
early! Naturally, prevention is an on-going effort. It
is the key to success! Dr. Carter expands on this in
the 'Development Cycle' of establishing the program.

PHASE TWO: CONTROL: You must have 'Standards of Workmanship for
Quality' and a formalized 'Hands-on Training Program'
to qualify your people to perform required 'Special
Processes'. Your package can consist of Drawings, Specifications, Design Standards, Visual Workmanship Standards, etc., that you can control your products & people
to. Control 'Goals & Objectives' must be set or you will
be consistently inconsistent and out of control. Process
Control Charts & Statistical Quality Control Techniques
will provide visual evidence & facts from which good engineering & management decisions can be made 'Quickly,
Quietly & Cost Effectively'. You must be 'In Command and
Control 365 days a year'. If you will follow the program
and the system, control will be there. Quality Control
Circle Activity by that or other names is included. However, documented & verbal commitment by management is required for life-long return on investment.

PHASE THREE: ASSURANCE: When the Prevention & Control Phases are in
place and functioning, 'Assurance Check Points' will allow you to check & assure that your Products, Processes,
Procedures, Program, Services & People are performing.
Closed-loop Assurance Feed-back will pay-off every-time!
This results in communication, corrective action & customer satisfaction, which brings new & repeat business. The
'Assurance of Quality in Product & Service Industries' is
and will continue to be the difference between success &
failure. Assurance is mandatory, not desired.

PHASE FOUR: ASSISTANCE: This requires the complete development &
                  implementation of finely tuned, documented policies &
                  procedures to allow you to 'Assist Employees' in per-
                  forming their tasks & job assignments, right, the first
                  time. These will also allow you to assist all other de-
                  partments with whom you interface & interact in the com-
                  pany, in addition to assisting customers with problems.
                  These policies & procedures must provide the means, meth-
                  ods, and techniques for immediate reaction & corrective
                  action on your part to resolve & prevent problems. You
                  must be willing to 'Assist' those in need. You can't say
                  'That's not my job'. Management must learn to listen to
                  learn. Teamwork is a must for survival! Quality Assurance
                  must be a small professional management team, geared to
                  assisting and auditing. People to people performance.

PHASE FIVE: AUDITING: We must conduct unbiased Quality Assurance Aud-
                  its on a random or scheduled basis for assurance. You can
                  audit policies, procedures, programs, projects, products
                  and people as related to Quality, Reliability, Safety,
                  Training, Materials, Methods, Techniques, etc. An audit
                  is the lowest level of inspection requiring the lowest
                  amount of people you can put into a program while realiz-
                  ing the maximum benefit from it. When you get your checks
                  and balances in place, you can perform audits that will
                  tell you whether your total program is functioning as it
                  should be. Without an audit function, you are out of con-
                  trol and out of touch with reality. All technical & man-
                  agement audits must be performed randomly by highly quali-
                  fied quality assurance people. The goal is to improve the
                  quality of people, product & service to the point that you
                  have high levels of confidence provided by auditing.

COMPLETION: The positive approach is that the 'Implemented Five Phase
                  Quality Program' will allow your organization & company
                  to operate with the lowest level of 'People Power' neces-
                  sary and at the highest possible return on investment. As
                  this comes true, you will have the right people, the right
                  robot, and the right material in the right place at the
                  right time, for maximum yields, productivity & profit. The
                  teamwork will be there through caring, sharing, participa
                  tive management programs and the Quality Assurance Program
                  in concert with each other. Attitudes will change and you
                  will find motivated people motivating others. You will at-
                  tain maximum quality at the lowest total cost. You will be
                  at your lowest level of turn-over & lowest absenteeism be-
                  cause you will have a 'Motivated Workforce' who will put
                  out 110% for you because they understand the importance

of what they are doing; they will understand that you care about them as individuals who are contributing to the total team, to the goals & objectives of their organization & their company; they will understand that you respect them, their thoughts, their ideas, and that you will respond and recognize their contributions in good faith. As such, they will react, respond and respect you as a manager and as a team member. Quality Awareness is not really a 'Program'; It's good business and good management practices as practiced by professional managers who care. In truth, you will not survive without a 'Total Quality Assurance Program' with the ingredients as prescribed in this "FIVE PHASE QUALITY PROGRAM" by Dr. Carter. Management needs to understand that quality is not free, but that it does pay great dividends every day, 365 days a year.

FIVE C's FOR THE FIVE PHASE QUALITY PROGRAM

1. CONTINUITY: THE QUALITY PROGRAM MUST HAVE

   COMPLETE CONTINUITY TO MEET FUTURE NEEDS &

   FUTURE TECHNOLOGICAL DEVELOPMENTS......

2. COMPATIBILITY: PEOPLE AND PROGRAM!

3. COMMUNICATION: PEOPLE, PROGRAM, POLICIES &

   PROCEDURES FROM START TO FINISH.......

4. CONTROL: PRODUCT, PROCESSES, PEOPLE & PROGRAM

5. COST-EFFECTIVENESS: THE BOTTOM LINE! QUALITY,

   RELIABILITY, SAFETY & TOTAL MATERIALS MANAGE-

   MENT AT THE LOWEST TOTAL COST.

                                        Dr. Carter

How You Can Save with A Wood Stove

Stove, Pipe. Installation, Etc. ....$	458.00
Chain Saw	149.95
Gas And Maintenance For Chain Saw	44.60
4-Wheel Drive Pickup, Stripped	8,379.04
4-Wheel Drive Pickup, Maintenance	438.00
Replace Rear Window Of Pickup (Twice) ..	310.00
Fine For Cutting Unmarked Tree In State Forest	500.00
Fourteen Cases Michelob	126.00
Littering Fine	50.00
Tow Charge From Creek	50.00
Doctor's Fee For Removing Splinter From Eye	45.00
Safety Glasses	49.50
Emergency Room Treatment (Broken Toes-Dropped Log)	125.00
Safety Shoes	49.50
New Living Room Carpet	800.00
Paint Walls and Ceiling	110.00
Worcester Chimney Brush And Rods	45.00
Log Splitter	150.00
Fifteen Acre Woodlot	9,000.00
Replace Coffee Table (Chopped Up And Burned While Drunk)	75.00
Divorce Settlement	33,678.22
Total First Year's Costs $54,922.81	
Savings In "Conventional" Fuel-First Year	62.37
Net Cost Of First Year's Woodburning $54,815.44	

From: Allen American, Allen, Texas

Comment by Dr. Carter: "Some 'COST REDUCTION PROGRAMS' can get
                        very costly!"

94

<u>QUALITY, RELIABILITY & SAFETY AWARENESS PROGRAMS</u>

Any 'Quality Awareness' or 'Quality Circle' Program will
work and yield results if management will just make sure
to Administer the Program and be sure to adhere to their
own management system of prevention, control, assurance,
assistance and auditing with regard to QUALITY, RELIABIL-
ITY, SAFETY & TRAINING. The 'Quality Awareness Program'
will not work unless you have your Quality Program in
operation. Management must learn to PLAN and make their
People a part of the over-all plan. Management must really
understand that this is a LONG TERM PROGRAM of 3 to 5....
4 to 8....5 to 10 years or more as based on circumstances
within each specific company. This is not a 6 month effort
as so many management programs seem to be. When the program
does not fully pay-off in 6 months, management interest is
lost and the program fails as most all will do when you
only look for the short term gains and don't even think
about the long term benefits which are many. Here are 6
basics required for a successful 'Awareness Program'.

1. Trust your people. They will do great things if you do
   and hurt you bad if you don't.

2. Develop Loyality to the firm. Start with an Orientation Program the first day as a part of your Total
   Quality Program and Personal Career Development Program.

3. Train & Develop your Prime Assets which are your People.
   Training is Prevention, Control & Assurance and you
   can't beat it for flexibility & retention of people.

4. Care about your People and Recognize their contributions
   and efforts. Treat each worker as an Individual if you
   expect them to care about your company.

5. Voluntary activities must be stimulated. The people must
   want to or the program will not function. A management
   program that is forced will fail.

6. The work environment must be a place where people can think
   and use their wisdom, make decisions that affect them and
   be asked to participate in the decision making process. I
   call this Communication, Coordination & Cooperation and
   working together for the good of the cause which is Growth
   and Development of the Person and the Company.

                                        Dr. Carter

QUALITY, RELIABILITY, SAFETY & TRAINING
BENEFITS*

- documented & functioning Q.R.S. Program
- personnel involvement in problem solving
- prevention of problems via participation
- efficient control of processes & products
- audit of total Q.R.S. management system
- implementation assistance
- prioritized goals for planned growth
- reduced scrap, repairs & rework costs
- reduced quality costs
- improved communication, coordination & cooperation

- expedient, visible results
- concise, clear reporting systems
- trained, qualified, motivated, goal oriented personnel
- product liability risk reduction
- improved productivity & profitability
- long term career development of personnel
- identification & retention of talented people
- satisfied customers

*Commensurate with management interest, dedication & involvement.

YOU....

YOUR PEOPLE....

YOUR COMPANY....

YOUR CUSTOMERS....

EVERYONE BENEFITS....

COMMENSURATE WITH MANAGEMENT INTEREST, DEDICATION, INVOLVEMENT & SUPPORT....

Dr. Carter

PRACTICAL
PREVENTION CONCEPTS

The Prevention of Problems is Cost-Effectiveness at its best.
Prevent the Problems up-front and you will never need to deal
with them again...Prevent Problems & Reduce Re-work, Repairs,
Rejects, Material Review Actions, Vendor Errors & Costs!

* **COMMITMENT!**	To Total Quality Management; Quality Right The First Time; Quality Improvement Process; Trust your People.
* **PARTICIPATION!**	Involved Leadership; Quality Teams
* **CONSISTENCY!**	Design & Produce to Quality Standards; Consistent Quality Services.
* **COMMUNICATION!**	From top to bottom; Allloops closed; From People to all management levels
* **COORDINATION!**	Teamwork; Factual Data; No Crutches
* **COOPERATION!**	Partnerships for Quality; Quality Teams
* **CORRECTION!**	Corrective Action; Complete; Control
* **COMPLETION!**	On Time, Everytime per Workorder, Dwgs., Quality Standards, as agreed
* **CONTROL!**	Processes, Products, Procedures, Who did What, When, Where.....
* **AUDIT!**	Anything, Anywhere, Anytime & Document
* **ASSURANCE!**	Check it...Assure it is Right! Put your Name or Stamp on it! Test, Verify.
* **ACTION!**	Make things happen! Teamwork; Do It!
* **REDUNDANCY!**	Back-up; Must have for Reliability
* **REVIEW!**	Design; Data; Document
* **RECORD!**	Data; Details; Facts; Reports
* **REPORT!**	Problems; Suggestions; Recommendations; Facts; Figures; Details
* **PLAN!**	To Prevent, Control, Assure, Assist, Audit
* **TRAIN & DEVELOP!**	Every Person for Growth, Skill, Quality
* **FOLLOW-UP!**	Close Loops; Make Sure; Complete, Report
* **QUALITY!**	Assures Success & Customer Satisfaction

These are Typical. They will help in Preventing Problems. These
will help you think of others as you achieve Goals & Objectives.

"Procurement Quality Assurance & Materials Quality Assurance"

By Dr. C. L. (Chuck) Carter, Jr., P.E.

Introduction:

The need for good commercial suppliers is of primary concern to most every purchasing manager and buyer in business and industry today. Purchasing good commercial quality products on time at the lowest total cost is of primary interest and concern to all purchasing and quality management at all levels. The person responsible for the total materials management program in your firm is a vitally interested person since any quantity of failures will hinder the entire operation and immediately add costs that were not a part of the original purchase agreement.

A clear, concise communication document must be developed to assure all parties that the good commercial quality products you have ordered will be the quality you want, when you want them, and at the price you agreed to pay. With these thoughts in mind, let me suggest and review some of the documentation that will be needed to accomplish this task which is the normal responsibility of Engineering, Materials, and Quality Assurance Management.

Procurement Quality Assurance:

A Procurement/Vendor Quality Assurance Policy must be established to determine who does what, when, where, why and how. I would suggest the following for your consideration:

Policy:

Purchasing is responsible for buying Quality products and materials on time at the lowest total cost from qualified suppliers. Purchasing is the "Central Control Contact" with all vendors and no commitment is to be made to a vendor by anyone, except Purchasing. Purchasing is to be directly involved and copied in all contacts with vendors.

This Company will have an "Early Warning Vendor Quality Assurance Program" to provide timely and advanced and on-going information on our vendors in order to assist Purchasing (and/or Engineering on new sources) in obtaining Quality products, on time, at the lowest total cost. The Quality Assurance Department will conduct Vendor Surveys, Source Inspection (in the best interest of the Company) and perform Receiving Inspection on all items and materials which go into or form a part of the Company's products. Records will be maintained by Quality Assurance with a copy of each Vendor Evaluation given to Purchasing for decision-making purposes. Purchasing will also receive a copy of all Rejected Material Reports for transmission to the vendors. Purchasing will also receive a copy of the Quality Assurance Analysis of the Vendor History Records and any other information Quality Assurance can provide to assist Purchasing. These records and reports will also be supplied to Executive Management as a part of the Company's Management Information System for maximum communication; coordination; and awareness. Purchasing and Engineering are to avoid Single Source Procurement in the best interest of the Company with regard to Quality, Delivery, and Price.

1. Responsibilities

   A. Purchasing is responsible for including Quality Requirements on the purchase orders and contracts they negotiate with vendors. Quality Assurance will assist Purchasing. Purchasing is responsible for notifying Quality Assurance of the need for evaluating vendors when new sources or second sources are being considered for major items or long-range procurement or special process suppliers, etc. Purchasing and Quality Assurance will coordinate a schedule of proposed vendor evaluations and/or any meetings with vendors. Purchasing will make all arrangements with the vendor for the survey, meetings, i.e. dates, time, etc. Purchasing is the "Central Control Contact" with all vendors.

   B. Engineering is responsible for notifying Purchasing and Quality Assurance on all proposed new sources, so Vendor Survey schedules can be established well in advance of the need for the production procurement. Whenever possible, Engineering should locate at least two (2) qualified sources for all new items for multiple-source procurement in the best interest of the Company. Purchasing specifications are to be provided by Engineering on all High-Dollar items such as the Computer; Teletypewriters; Controllers; etc., and on all special procurements such as Acoustic Couplers; Special Process Items, such as Printed Circuit Boards; wire-wrapped assemblies; sheet metal parts; cabinets; and other items where a drawing or specification is needed to buy, inspect, test, or otherwise determine the acceptability of the product or material. Specific requirements for Quality, Reliability, Safety, Interchangeability, Maintainability, etc., shall be included in Purchase Specifications and Drawings.

   C. Quality Assurance will conduct Vendor Quality Surveys; perform Receiving Inspection; conduct Source Inspection; provide inputs into the Purchase Specifications and purchase orders; and maintain records of all related Vendor Quality activities which will be provided to Purchasing, Engineering, and other management people in the form of timely reports. Quality Assurance, Purchasing, Material Control, and Engineering will work together to resolve Vendor Quality problems through: meetings with vendors; Material Review Board activities relating to corrective action by the vendor and/or the Company; the need for new sources of supply or replacing problem vendors when extensive communication, coordination, and cooperation on the part of the Company has not resulted in a supplier that provides Quality products, on time, at the lowest total cost.

The next major item of interest will be the establishment of the Quality, Reliability, Safety and Related Requirements for the Purchase Specification. This will be the communication document the vendor will respond to on a Request for Quote; Request for Proposal; Purchase Order or formalized Contract from Purchasing.

You will recognize the words to be clear, concise and direct with regard to all areas that affect the Quality, Reliability, Safety, Maintainability, Interchangeability, Configuration, Documentation, Product Liability, and other important considerations.

<u>QUALITY ASSURANCE, RELIABILITY, & RELATED</u>
<u>REQUIREMENTS FOR ALL PURCHASE SPECIFICATIONS</u>

Scope: These Requirements are both general & specific with regard to Vendor Quality Assurance Systems Management, Reliability, Interchangeability, Maintainability, Safety, Configuration & Change Control, Documentation, Inspection & Test, Records, Packaging, Shipping, etc. These are to be a part of every Purchase Specification. The Technical Engineering/Special Procurement Requirements will be done by Purchasing & Engineering Departments as may be necessary.

1.0   <u>Quality Assurance & Related Requirements</u>

   1.1 <u>Vendor Quality Assurance System/Program</u>: The Vendor will have & maintain a documented & functioning Quality System for the Control & Assurance of Quality, Reliability, Safety, Interchangeability, Maintainability, etc. as stated in SCOPE above. The Vendor's Quality System and all records will be subject to a Formal Survey/Evaluation and or Audits as conducted by our Quality Assurance Department as scheduled & coordinated thru the Purchasing Dept. The buying company recommends and will use ISO/ANSI/ASQC Quality Standards of the 9000/Q90 Series as the Standards for Evaluating the Vendor and his Quality Program. The Buying Company uses these Standards internally and recommends either series(which are identical) to its Vendors. It is the intent of the Buying Company to require all of its Vendors to enter into its Vendor Certification Program which includes 'JUST-IN-TIME' and 'DOCK-TO-STOCK' concepts and philosophies which will benefit both the Vendor and the Buying Company.

      1.1.1 Vendor questions relating to any of the above or those regarding the Purchase Order or Purch. Spec. should be directed to the Purchasing Dept. for reply. Purchasing will coordinate the questions with Engineering, Quality Assurance, etc., and reply to Vendors in a timely manner.

   1.2 <u>Inspection & Performance Testing</u>: The Vendor is responsible for the performance of all Inspection & Testing as required to meet or exceed the requirements of this Purchase Spec., Purchase Order, and all associated drawings. Each item will initially be Inspected & Tested by the Vendor with records maintained on each, until sufficient Quality History of acceptable products permits Sampling, Reduced Sampling, Skip-lot, etc. The Buying Company will maintain their Reporting System and advise the Vendor of its status, including Qualification & Certification for J.I.T. & D.T.S. programs.

      1.2.1 <u>Inspection on Receipt</u>: The Buyer reserves the right to perform any Inspection or tests to assure conformance to this spec. and to reject any item that does not conform for cause.

      1.2.2 <u>Source Inspection</u>: The Buyer may perform S.I. and or Audit/Monitor the Inspection and

100

Testing operations at the Vendor's facility. All Source
Inspection, Audits, or other evaluations at the Vendor's
facility will be arranged through the Buyer's Purchasing
Department.

1.2.3 <u>Quality Assurance Records</u>: Quality Assurance records covering the item/product shall be retained on file at the vendor's
facility for a period of one year (unless otherwise extended)
and shall be made available immediately upon request.

1.3 <u>Quality, Reliability, Interchangeability, and Maintainability</u>: The
Quality, Reliability, Interchangeability, and Maintainability of
the designed item shall be of the highest caliber. All parts, assemblies, subassemblies, components, etc., shall be designed with materials of High Quality and Reliability. Interchanging one item with
an item having the same part number must produce equivalent or
improved reliability/results and shall require no rework or modification to install. The finished item/product shall be delivered
free of defects and shall be reliable and maintainable with minimum
downtime; minimum tools; no special tools; minimum training of personnel; and no major disassembly for maintenance or repair. Mean
time between failures (MTBF) and mean time to repair (MTTR) shall
be as stated on the Engineering Drawing/Specification, if required.

1.4 <u>Configuration Control and Identification</u>: Each item/product and
part shall be identified with the Buyer's part number and revision
code. Each cabinet or chassis shall be identified (label, tag,
stamp) with the applicable numerical information and serial number
for traceability, configuration control, and recall capability.

1.4.1 <u>Documentation and Reproduction</u>: Documentation, final test
and inspection procedures, operation and maintenance manuals
shall be provided to the Buyer with each item. The Buyer
shall have reproduction rights to the documentation as provided by the Vendor for the Buyer's in-plant testing and
inspection, training, field service, and customer use.

1.4.2 <u>Product Changes: Incorporation, Notice, etc.</u> The Vendor shall
have the responsibility to assure the Buyer that any changes
to the item/product that affects Form, Fit, Function, Appearance, or Performance, in any way, shall not be initiated
without prior written approval from the Purchasing Department
Manager. The Manager of Purchasing will coordinate the necessary approvals from Engineering and Quality Assurance. It
shall be the responsibility of the Vendor to advise Purchasing
Management in writing, of the effectivity by unit, part, product, and serial number, in which the changes will be incorporated, after formal written approval by the Buyer to the Vendor.
The Buyers approval of such change shall be presumed, unless
the Vendor is notified in writing by Purchasing Management
within thirty (30) days of receipt of the notice of such
change, except in cases of mutually agreed written extensions.
Any change rejected by the Buyer shall not be incorporated
into the Buyer's equipment. If design changes should be
necessitated by the failure of a Vendor product to meet the
requirements of this specification, purchase order or related

drawings, the Vendor agrees to either retrofit parts or provide retrofit parts and installation procedures, at the Buyer's option, for all items products, parts delivered with such design defects, at no cost to the Buyer.

1.5 Safety: The item/product shall be safe. The Vendor shall comply with the Occupational Safety and Health Act of 1970 and all other laws and executive orders related to Product, Consumer, and/or Personnel Safety.

    1.5.1 Underwriters' Listing: The item/product must be listed or formally (documented) in process of investigation for listing by the Underwriters' Laboratories, Inc. The item/product shall be identified as required by U.L. and the Vendor shall maintain the product and production processes to the satisfaction of U.L. to maintain the U.L. Listing.

1.6 Product Liability: The Vendor shall defend and hold harmless the Buyer (by company name) from any and all claims, demands, causes of action, damages, costs, expenses, losses or liabilities, or injury to or death of any person, and damage to or loss or destruction of any property (including, but not limited to, the Buyer's property or the property of its customers), arising out of Vendor's performance or out of the performance of the Vendor's product. The Vendor is liable for his product and for the performance of his product and for any and all damages and losses caused as the result of his product.

1.7 Packaging and Shipping: The Vendor is responsible for assuring that all items/products shipped to the Buyer or drop-shipped to the Buyer's Customer Site, will be packaged to prevent in-transit or handling damage of any kind. The Vendor shall provide a copy of their Packaging Procedure to the Buyer as a part of this purchase specification. Packaging problems and damage to the item/product will require the Vendor to take immediate corrective action in writing to the Director of Quality Assurance and the Manager of Purchasing. Each shipping container shall be marked: "Fragile, Electronic Parts, Do Not Drop, (or equivalent words of caution)."

    1.7.1 Shipping: The Buyer reserves the right to determine the carrier or method of transportation to the Buyer's Plant or any other location.

    1.7.2 The Vendor shall conduct a Wrap, Pack, Packaging and Shipping Inspection, prior to releasing the item/product from the Vendor's facility. If the item/product is packaged for shipment at an outside packaging supplier, the Vendor shall be responsible for all of the above.

Conclusion:

Today, the Consumer, Commercial and Industrial Material marketplace requires joint protection for the Vendor and the Vendee with regard to all elements of Quality, Reliability, Safety, Maintainability, Interchangeability, Configuration, Change Control, Documentation, Inspection, Test, Records, Packaging, Shipping and all the rest. You need to tell it the way it is and be sure the Vendor understands all the critical aspects of the purchase specification and the need for compliance. When in doubt, spell it out!

As you continue to learn about your need to comply with the many Safety Regulations and Laws, and as you recognize your involvement or potential involvement in Product Liability and Related Warranty lawsuits, I suggest and recommend that you develop a clear, concise Purchase Specification and cover all aspects of Quality, Reliability and Safety that will allow you to Purchase Good Commercial Quality Products and Services in the future, with confidence.

HOW TO CONTROL THE QUALITY OF YOUR QUALITY CONTROL.....

MANAGEMENT MUST COMMIT TO CONSISTENT, RELIABLE QUALITY ASSURANCE, THAT IS DEDICATED TO PREVENTION, CONTROL, ASSURANCE AND AUDITING......

Dr. C. L.(Chuck)Carter Jr., P.E.

Nine Ways That Quality Companies Produce Safe,
Reliable, Quality Products, Profitably

1. They Build Quality Attitudes....and Insist on Safety!

2. They Select Qualified Operators....and Instruct Thoroughly!

3. They Make Sure Operators UNDERSTAND What is Expected....and
   Then They Follow-up to be Certain the Instructions are Followed.

4. They Explain to the Operators What Caused The Rejects....and
   How To Correct The Defects in the Future.

5. They Compliment Operators On Quality Work.....and **Make** Sure
   That Safety Procedures are Followed.

6. They Communicate & Cooperate With ALL Departments to Help
   Correct Causes and Solve Problems.

7. They Have A Planned Program for Contacting and Selling ALL
   Employees on Quality, Reliability, Safety, Service & Integrity.

8. They Explain How ALL Employees Benefit by Producing Quality
   Products, On Time, At The Lowest Total Cost to the Company.

9. They Encourage Operators to Submit Their Ideas on Improving
   Quality, Safety and Product Integrity.

   I am sure that you can add to this brief list and capitalize
   through total involvement in your company....

10.

11.

12.

13.

14.

15.

TYPICAL QUALITY - RELIABILITY PROBLEMS
RESULTING FROM RELATIONS WITH ASSOCIATED DEPARTMENTS

Engineering:

1. Language, communications and understanding
2. Specification and Drawing Tolerances too tight
3. Change Orders - General Documentation
4. Lip Service to Quality and Reliability
5. Lack Understanding - Units that work but do not meet Requirements are not Acceptable
6. Faulty Analysis of Problems

Purchasing:

1. Buying from Unqualified/Non-Approved Vendors
2. Language, communications, coordination, understanding
3. Heavy emphasis on initial price and delivery - light on Quality
4. Single Source Vendors
5. Vendor Evaluation, Rating and Analysis

Contract Administration:

1. Language, communication, understanding
2. Fail to Read and Understand Contract Before Signing
3. Contract Change Orders - Coordination

Manufacturing:

1. Schedule over rules Quality
2. Unwilling to Accept Responsibility for Quality
3. In process controls at all levels of manufacturing and not just Final Inspection
4. Control of Discrepant Material
5. Inadequate Failure Data for Corrective Action
6. Communications, coordination, cooperation

"THE DEVELOPMENT OF A QUALITY ASSURANCE ORGANIZATION"

Management must first make the commitment to Quality, Reliability, Safety & Training. If management has made that commitment to Prevention, Control, Assurance, Assistance and Auditing, and has actually documented a formal 'Quality Policy' for achieving Product & Service Quality, you still have the major task of developing the Quality Assurance Organization. This will be the organization that will evaluate, analyze, generate, document and implement the Total Quality Assurance Program.

The major question asked by most people is "Where should the Quality Assurance function report?" I feel it should report to executive management. Others may say it should report to manufacturing or engineering. I have seen it work in many organizations and fail in many others. The necessary ingredient is a 'Good Blend' of professional talent. You must have qualified responsible people in each of the key management positions. Quality is a Key Management Function. You must have clear individual & departmental responsibilities spelled out for good communication and understanding. Executive management must monitor ALL of their functions via 'Operations Audits' to maintain a constant awareness and not let themselves get into an 'Out of Control Status' with regard to their management team.

Some of the specifics that management must keep in mind are:

1. The Quality Assurance Manager or Director must be a professional person and recognized as a member of the 'Executive Management Team'. This must be a Leader!

2. The Quality, Reliability & Safety Organization must have trained & technically competent people who can relate to all levels of management & personnel in every department. Participative, not Autocratic!

3. Professionalism is a must if the company is to succeed. Quality Control can be a part of Quality Assurance or Q. C. can be an in-line function of manufacturing, but in either case Q. C. is much more than 'Inspection'.

4. The Q. A. function must be professionally staffed with flexible people who can work with sales, engineering, production, accounting, purchasing, customers, etc. Acceptance of Product is based on 'Facts' not opinion. Achieving Quality & Safety is no easy process.

Dr. C. L. (Chuck) Carter Jr., P.E.,

Does Your Team Pass The Teamwork Quiz?

In a difficult economy — after you've reduced your staff to a minimum — maximum productivity and profit may depend upon **team effectiveness.**

Contrast the *productive value* of a highly committed team which communicates, coordinates, speaks its mind and pushes for new, more and better, against the *productive value* of a team which fails to operate in this way.

The following quiz may help you evaluate whether your team has the qualities commonly observed on effective teams.

Yes	No	Do the members of your team:
___	___	1. Know, and feel committed to, team goals?
___	___	2. Problem-solve in direct and practical ways?
___	___	3. Participate fully?
___	___	4. Express their ideas freely?
___	___	5. Seek new and better ideas and methods?
___	___	6. Stay involved, interested and committed?
___	___	7. Listen to each other actively and effectively?
___	___	8. Take the risk of being open and honest?
___	___	9. Disagree freely and openly?
___	___	10. Respect and explore differences of opinion?
___	___	11. Ask for and give honest feedback?
___	___	12. Handle conflict openly and constructively?
___	___	13. Express real feelings as well as ideas?
___	___	14. Remain aware of the impact of their own behavior?
___	___	15. Periodically review how they are working together?

If you rated your team as having more "Yes's" than "No's", you probably already have a highly effective team.

"TEAMWORK IS THE ONLY WAY TO ACHIEVE QUALITY AND RELIABILITY"

TAKE THE TIME TO PLAN ....

THEN FOLLOW THE PLAN ....

WE MUST LEARN TO LISTEN TO LEARN TO LEAD ....
WE MUST LEARN THAT WE MUST PLAN FOR THE LONG TERM ....
WE MUST LEARN THAT OUR FUTURE IS "QUALITY & RELIABILITY"
WE MUST LEARN THAT WE CAN ONLY MAKE IT THROUGH R & D ....
WE MUST LEARN TO INVEST IN OUR PEOPLE, NOW, IF WE
PLAN TO BE AROUND IN THE FUTURE ....

DR. CARTER

Practise.

I do it normal
I do it slow
You do it with me
Then off you go

FOOD FOR THOUGHT

I KNOW YOU BELIEVE YOU UNDERSTAND WHAT
YOU THINK I SAID, BUT I AM NOT SURE YOU REALIZE
THAT WHAT YOU HEARD IS NOT WHAT I MEANT.

"TAKE TIME TO PLAN"

Here are 10 specifics that will help you to get there
from where ever you are.....

1. Take Time To Work....It is the price of Success

2. Take Time To Think.....It is the source of Power

3. Take Time To Play.....It is the secret of Youth

4. Take Time To Read.....It is the fountain of Knowledge

5. Take Time To Worship.....It washes the dust of earth from
                           our Eyes & Allows us to see Truth

6. Take Time To Love.....It is the sacrament of Life

7. Take Time To Enjoy Friends.....It is the source of Happiness

8. Take Time To Dream.....It is the road to Greater Vision

9. Take Time To Laugh.....It is the music of the Soul

10. Take Time To Plan.....It is the secret of being able to have
                         Time to Take Time for the first nine!

Ref: AMS - Dallas
     Dr. Carter

TIME....TIME...TIME...

                    WE ALWAYS NEED MORE.....
                    BUT THERE IS NO MORE THAN:

        24 Hours a Day......

        168 Hours a Week.......

        8760 Hours a Year........

THAT'S ALL THE TIME THERE IS....and WE ALL START OUT
WITH THE SAME AMOUNT! WE ALL START OUT EQUAL!

HOW WELL WE MANAGE OUR TIME, WILL DETERMINE HOW MUCH
WE ACCOMPLISH.

THOSE WHO ACCOMPLISH THE MOST, HAVE LEARNED TO MASTER
ALL THE TIME WE HAVE......

THEY PLAN....THEY SET GOALS & OBJECTIVES and THEN THEY
FOLLOW THEIR PLANS.

IN EFFECT, THEY PREVENT, CONTROL, ASSURE, ASSIST & AUDIT!

IN EFFECT, THEY HAVE A PERSONAL QUALITY ASSURANCE PROGRAM!

IN EFFECT, THEY HAVE A PERSONAL CAREER DEVELOPMENT PROGRAM!

IN EFFECT, THEY HAVE AN ON-GOING TRAINING & DEVELOPMENT
PROGRAM IN ACTUAL OPERATION.

REALISTICALLY, THESE INDIVIDUALS WILL GET TO WHERE THEY
ARE GOING, IN THE SHORTEST PERIOD OF TIME....

                    Dr. Carter

CAREER DEVELOPMENT.....

THE HUMAN FACTORS & THE QUALITY OF WORK LIFE

Quality of Work Life has been identified as "The process
used by an organization to unlock the creative potential
of its people by involving them in decisions affecting
their work lives" (R. Guest, Harvard Business Review, of
July, 1979). Ref: ASQC/ASTD/AMS/SME & other Prof. Societies.

The concept provides an opportunity to maximize human resources and reverse the trend toward declining productivity and more government regulations. In recent years the
concept has grown into an active movement within some areas
of U.S. industry in both large and small companies. It must
spread into all segments of business, industry & government
to allow us to continue to compete in the markets of the
world. Another term I use with regard to the subject, is
'CAREER DEVELOPMENT'. My paper: "CAREER DEVELOPMENT, WHAT
ARE YOU DOING...WHERE ARE YOU GOING?" as presented at the
American Society for Quality Control Conference in Atlanta
in May 1980, was very well received because it was and is
very personal to every person in every company. Some of
your people were there. They listened, heard, and were impressed and touched because it addressed their personal
needs in the world of work. Likewise, I was impressed with
the reception and acceptance of my presentation. One statement I made went something like this: "Very few companies
in the United States at this time have a Career Development
Program. Some talk a good game, make some promises, but do
nothing. As a result, management is causing and creating
costly turn-over of their most vital resource. Some change
the name of their Personnel Department to the Human Resource
Department. This accomplishes nothing unless the company
adopts and implements a Career Development Program that will
really involve, train, develop and consider the whole person
as a resource and as a gold mine with a golden mind worth
capturing until retirement. I further stated that a Career
Development Program would become a "STANDARD" in the business
and industrial world because I don't think business & industry
will be around in the future unless they have People Oriented
Career Development Programs."

We must 'Train Trainers to Qualify the Un-qualified Production
Technician to produce Quality Products, On Time, at the Lowest
Total Cost'....We must free managers to manage more....We must
Prevent, Control, Assure, Assist & Audit....We must make the
person and the department responsible for the quality of that
which they do; We must take away the crutches and get everyone
to accept their responsibility for Quality, Reliability & Safety;
We must Prevent Problems; We must, if we are going to compete.

                                              Dr. Carter

AS A MANAGER.... AS A PROSPECTIVE MANAGER....

YOUR GREATEST PROBLEM WILL BE/IS.....

Employee
Problems...

"QUALITY MANAGEMENT SEMINAR" By: Dr. Carter

Increased Productivity Through Mechanization

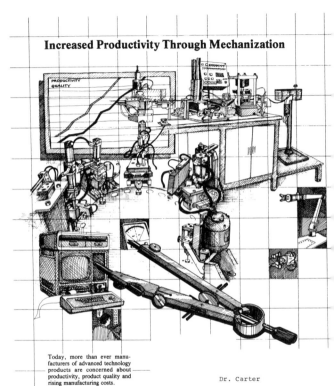

Today, more than ever manufacturers of advanced technology products are concerned about productivity, product quality and rising manufacturing costs.

Dr. Carter

YOU, AS THE INSPECTOR TODAY, MUST BE CONCERNED ABOUT THE QUALITY OF YOUR EDUCATION, TRAINING & KNOWLEDGE FOR THE FUTURE. WHAT YOU DO AND HOW YOU DO IT TODAY WILL BE MUCH DIFFERENT IN THE FUTURE! YOU MUST MAKE YOURSELF READY FOR THE FUTURE...BUT YOU MUST START TODAY! EVALUATE YOUR CAREER PLANS TODAY AND GET READY FOR THE FUTURE IN AUTOMATION...ROBOTICS...COMPUTERIZATION. GET READY!!

BEATEN BEFORE YOU START

If you think you're beaten, you are...
   If you think you dare not, you don't....
If you'd like to win but think you can't
   It's almost a cinch you won't.
If you think you'll lose, you're lost,
   For out in the world you find
Success begins with a fellow's faith,
   It's all in the state of mind.
Full many a race is lost
   Ere ever a step is run;
And many a coward fails
   Ere ever his work's begun.
Think big and your deeds will grow,
   Think small and you'll fall behind;
Believe you can and you will
   It's all in the state of mind.
If you think you're outclassed, you are;
   You've got to think high to rise,
You've got to believe in yourself before
   You can ever win a prize.
Life's battles don't always go
   To the stronger or faster man,
But soon or late, the one who wins
   Is the one who thinks he can.

Author Unknown

Comment by Dr. Carter: Persistence, Determination & Planning
are all powerful. If you plan well &
follow the plan; understand yourself
as an individual; know your strengths
& capitalize on them; set realistic
goals & objectives that are achieve-
able & attainable and then get out
there and get after them on a steady
hour by hour & day by day basis...You
can get to where you are going in the
shortest period of time! 'Press On'!!
And refer to my 'Take Time To Plan'as
10 specifics to help you make it happen!

115

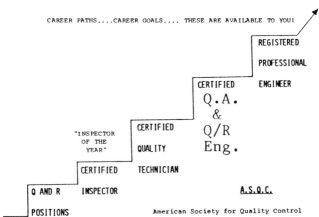

CAREER PATHS....CAREER GOALS.... THESE ARE AVAILABLE TO YOU!

REGISTERED PROFESSIONAL ENGINEER

CERTIFIED Q.A. & Q/R Eng.

CERTIFIED QUALITY TECHNICIAN

"INSPECTOR OF THE YEAR"

CERTIFIED INSPECTOR

Q AND R POSITIONS

A.S.Q.C.

American Society for Quality Control

Get Involved! Help Yourself to Grow & Develop! Its never too late! Dr. Carter Passed the Certified Quality Auditor Exam on 6/2/90. Follow the Career Paths & Go All The Way Up.... You Can Do It!

You & Your Career
Dr. C. L. Carter Jr., P.E.

1990

CAREER PATHS....CAREER GOALS.... THESE ARE AVAILABLE TO YOU!

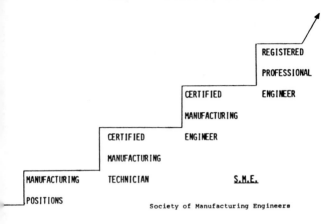

MANUFACTURING
POSITIONS

CERTIFIED
MANUFACTURING
TECHNICIAN

CERTIFIED
MANUFACTURING
ENGINEER

CERTIFIED
MANUFACTURING
ENGINEER

REGISTERED
PROFESSIONAL
ENGINEER

S.M.E.

Society of Manufacturing Engineers

You & Your Career
Dr. C. L. Carter Jr., P.E.

CAREER PATHS....CAREER GOALS.... THESE ARE ALL AVAILABLE TO YOU!

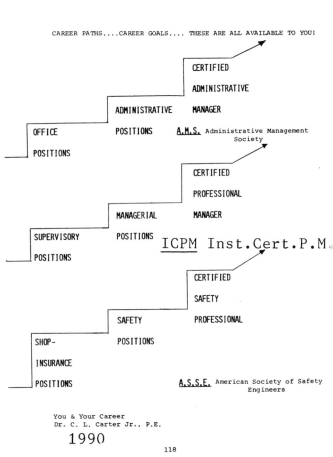

CERTIFIED

ADMINISTRATIVE

MANAGER

A.M.S. Administrative Management
Society

ADMINISTRATIVE

POSITIONS

OFFICE

POSITIONS

CERTIFIED

PROFESSIONAL

MANAGER

ICPM Inst.Cert.P.M.

MANAGERIAL

POSITIONS

SUPERVISORY

POSITIONS

CERTIFIED

SAFETY

PROFESSIONAL

SAFETY

POSITIONS

SHOP-

INSURANCE

POSITIONS

A.S.S.E. American Society of Safety
Engineers

You & Your Career
Dr. C. L. Carter Jr., P.E.
1990

<u>TEN VALUABLE TRAITS THAT CAN HELP GET YOU TO THE TOP</u>

Top Performers:

1. Have a Purpose in Life

2. Formulate Plans to Accomplish their Goals

3. Don't get Trapped in a comfortable 'Plateau' (stage of life) very long

4. Take Risks after determining what the consequences will be

5. Base their Self-confidence on History: Past Successes were a result of their Skills

6. Solve Problems rather than placing Blame

7. Rehearse Future Events Mentally...Always with a Positive Outcome

8. Like to Take Control

9. Are Concerned With Quality Performance, not just quantity

10. Train & Utilize those Around Them

Based on an extensive study over 16 years by the Peak Performance Center in Berkeley, Calif.

I would agree as based on my professional Career Counseling Experience over the past 19 years.....Dr. Carter

GUIDELINES FOR LEADERSHIP DEVELOPMENT......

Typical Leadership Traits: INTEGRITY, JUDGEMENT, LOYALTY, TACT, DIPLOMACY, CONDUCT, COURAGE, DEPENDABILITY, DECISIVENESS, ENDURANCE, ENTHUSIASM, INITIATIVE, KNOWLEDGE, UNSELFISHNESS, PERSEVERANCE, COMMITMENT, INVOLVEMENT, PARTICIPATION, LISTENS, RECOGNIZES, CARES, CONCERN, COMMUNICATES, COMMUNICATES, COORDINATES, COOPERATES, COUNSELS, APPRECIATES, UNDERSTANDS PEOPLE....

Eleven Basic Principles:

1. KNOW YOURSELF & HAVE A PERSONAL TRAINING & DEVELOPMENT PROGRAM

2. KNOW YOUR PEOPLE & TAKE CARE OF THEM....BE A LISTENER

3. COMMUNICATE CLEARLY WITH EVERYONE, OFTEN! BE PATIENT...

4. LEAD THE WAY...SET A GOOD EXAMPLE FOR YOUR PEOPLE, EVERYDAY!

5. TRAIN, EDUCATE, MOTIVATE & TEACH YOUR PEOPLE...ASSURE UNDERSTANDING, THEN AUDIT & FOLLOW-UP TO ASSURE COMPLETION ON TIME

6. DEVELOP TEAMWORK...THE "TOTAL QUALITY TEAM CONCEPT"

7. MAKE CONSISTENT DECISIONS IN A TIMELY MANNER

8. PROVIDE RESPONSIBILITY & AUTHORITY....DEVELOP LEADERS!

9. PLAN, ORGANIZE & COORDINATE YOUR HUMAN RESOURCES TO UTILIZE, GROW, DEVELOP & BUILD ON THEIR ABILITIES...REMEMBER, YOU SHARE 50% OF THE "CAREER DEVELOPMENT STREET"

10. MAINTAIN YOUR OWN TECHNICAL PROFICIENCY...NEVER STOP LEARNING

11. YOU ARE RESPONSIBLE FOR YOUR ACTIONS AND FOR THOSE OF YOUR PEOPLE AT ALL LEVELS....LEAD, GUIDE, DIRECT & DEVELOP YOUR GREATEST ASSET, YOU AND YOUR PEOPLE! BE A "CAREER COUNSELOR" AND "HELP THEM TO GET TO WHERE THEY ARE GOING IN THE SHORTEST PERIOD OF TIME.

Dr. C. L. (Chuck) Carter Jr., P.E.

How To

Cope

With

STRESS...
   STRAIN...
      CONFUSION...
         "SHIP IT PHILOSOPHY"

DEVELOP & IMPLEMENT A TOTAL QUALITY, RELIABILITY, SAFETY & TOTAL
MATERIALS MANAGEMENT PROGRAM WITHIN YOUR COMPANY TO ASSURE YOUR FUTURE
IN THE MARKETPLACES OF THE WORLD....

UNTIL YOU DO, THE STRESS & STRAIN WILL JUST GET WORSE!! WHY SUFFER?

Dr. Carter

TOTAL QUALITY MANAGEMENT & TOTAL QUALITY ASSURANCE IN THE NINETIES__
(The Control & Assurance of Quality, Reliability & Safety By Carter)

As we look at the future, 'The Era of the 90's' spells out the follow-
ing considerations for **QUALITY, HUMAN RESOURCES, ENGINEERING, MANUFAC-
TURING, SALES & MARKETING, CUSTOMER SERVICE, MANAGEMENT & ALL ASSOCIA-
TED FUNCTIONS:**

* **N - NOW** is the time for Executive Management to Commit to T.Q.M.
* **I - INVEST** in & Integrate All Human Resources for Quality Assurance.
* **N - NO** more Non-Conforming Materials. Zero is the Goal.
* **E - ENGINEERING** that is Quality, Reliability & Safety Oriented.
* **T - TOTAL** Quality Management Systems that are Fully Operational.
* **I - INTERNAL** & External Partnerships with People & Suppliers.
* **E - EVALUATION**, Efficiency & Emphasis on Certification by Auditing.
* **S - SAVINGS** & Profitability from People, Management, Suppliers, etc.
  In addition, the 90's will show that QUALITY CONTROL, QUALITY ASSUR-
  ANCE, MATERIALS MANAGEMENT, JUST-IN-TIME, DOCK-TO-STOCK, S.P.C. and
  CUSTOMER SATISFACTION WILL BE THE MAJOR FACTORS FOR SUCCESS....

In translating the considerations of the 90's with regard to Quality
Assurance in Manufacturing & Service Industries, it can be concluded
that the following factors are essential to "AN OVERALL MANAGEMENT
SYSTEM OF PREVENTION, CONTROL, ASSURANCE, ASSISTANCE & AUDITING."

1. The Total Q.A. & Total Materials Management Concept of Operation
   Must & Will Prevail or...You will be going out of business!

2. Customer/Supplier Relationships Must be Partnerships to Assure Success.

3. Continuous Control & Improvements of All Processes, Products/Services.

4. A 'Closed Loop System' which provides for 'Continuous Automatic Sampling' of In-Process Materials with Automatic Sample Analysis followed
   by Automatic Process Control Adjustments.

5. On-going incorporation of Automatic Test & Measuring Equipment into
   the Manufacturing Sequence of Operation to Assure Quality.

6. An Automatic Process & Q.C. System for the Continuous Evaluation
   of Selected Tooling, etc., used in the Manufacturing Cycle.

7. Mandatory 'Design Reviews' of Engineering's Progress concerning
   Quality, Reliability, Safety, Service & Product Design Integrity at
   Pre-Established Points as conducted by Executive Management or 3rd.
   party Management Consultants.

8. All Production People, Quality Inspectors/Auditors & Customer Service
   Technicians or Individuals who interface with Customers MUST BE TRAIN-
   ED, QUALIFIED & CERTIFIED to perform in any way on or about the Pro-
   duct(s). This also includes Engineering Technicians & Field/Customer
   Service Installation People!

9. An Operational Calibration System/Program per International Standards
   covering ALL Test, Inspection, Process Control & Engineering Tools,
   Gages & Equipment.

10. An Operational Supplier Qualification & Certification Program to
    Assure Achievement of J.I.T & D.T.S. in Every Case!

11. On-going Audits by Certified Quality Auditors, Any-where, Any-time!

THE INTEGRATED QUALITY MANUFACTURING SYSTEM
AS PROJECTED BY Dr. C. L. Carter, Jr.

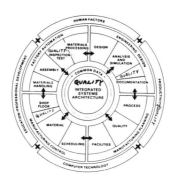

The "CASA Wheel" represents the basic model of a
computer integrated manufacturing system as it
would actually function in industry. By adding or
deleting functions, the wheel provides a framework
to plan the future implementation of modules into a
company's total integration plan.

C.A.S.A. = COMPUTER & AUTOMATED SYSTEMS ASSOCIATION OF S.M.E.

C.A.S.A. WHEEL AS MODIFIED BY Dr. C. L. (Chuck)Carter Jr., P.E.

QUALITY ... OF THE FUTURE - IS THE FUTURE

1. EVERY PERSON "A QUALITY PERSON".
2. EVERY INDIVIDUAL TRAINED IN Q.R.S.
3. QUALITY WITHIN EVERY DEPARTMENT.
4. QUALITY ASSURANCE PREVENTS, ASSISTS, AUDITS, REPORTS, CLOSES LOOPS, FOLLOWS-UP.
5. QUALITY COSTS WILL BE LESS THAN 1% OF SALES. (TODAY IT GOES TO 40% AND THAT IS PURELY A MANAGEMENT PROBLEM, SELF-INFLICTED, CAUSED AND CREATED BY MANAGEMENT.)
6. QUALITY MANAGEMENT SYSTEMS AND PROGRAMS WILL CORRECT THE CURRENT MANAGEMENT PROBLEM RATIO FROM 90% (BEING MANAGEMENT PROBLEMS) TO 10%.
7. QUALITY ... THE MONEY-MAKER. QUALITY SELLS AND SEALS YOUR MARKET SHARE FOR THE FUTURE.
8. QUALITY IS YOUR ONLY HOPE FOR THE FUTURE!

Dr. Carter

Dr. C. L. (Chuck) Carter Jr., P.E., C.Q.E., Projects:

"THE FUNCTIONAL QUALITY ORGANIZATION OF THE FUTURE"

```
┌─────────────────────┐
│  CHAIRMAN OF BOARD  │
│         OR          │
│  PRESIDENT & C.E.O. │
└─────────────────────┘
          ↕
┌─────────────────────┐
│  VICE PRESIDENT OF  │
│  QUALITY ASSURANCE  │
│         &           │
│  HUMAN RESOURCES    │
└─────────────────────┘
          ↓
```

TYPICAL FUNCTIONS:

1. Direct Lines to Every Function in the Company
2. Bring Human Resources In: Right Person, Right Career, Job/Concept
3. Orientation of All New Hires & Current People: Company, Products/People
4. Train, Develop, Motivate at All Levels, From Start Date to Retirement
5. Interface & Interact with All Departments
6. Assist People to Perform Consistently, Every Day in Every Way
7. Recognize, Care, Share, Communicate, Coordinate, Cooperate, Reward
8. Develop, Document & Implement the 'Total Quality Assurance Management System of Prevention, Control, Assurance, Assistance & Auditing'
9. Educate, Train & Sell the T.Q.A. Management System to Everyone
10. Prevent Problems with the Early Warning T.Q.A. Management System
11. Develop & Implement the 'Design Review' along with 'Program Evaluation Review Technique' (P.E.R.T.) for Constant & Consistent Administrative Awareness & Reporting
12. Control to Established Quality Assurance Policies, Procedures, and the Standards, Test Equipment, Product & Software Test Specifications
13. Assure via Check & Balance Quality Assurance Methodology
14. Integrate Quality Control into All Line & Staff/Administrative Areas
15. Audit Policies, Procedures, Projects, Programs, Products, People... Anywhere, Anytime, under Any Circumstances, to Prevent & Assure
16. Document & Report Every Audit to Executive Management for Action
17. Make Presentations to Customers & Prospective Customers: Audit Sites
18. Executive Member of the Profit & Productivity Team
19. Quality, On Time, At Lowest Total Cost & Highest Return on Investment
20. Quality People Produce Quality Products & Perform Quality Services for Quality Customers, Resulting in New & Add-on Business: Closed-loop Reference to the Company's Quality & Human Resources Policies & Procedures Manuals as Signed By Executive Management.

As Projected & Documented By Dr. Carter over the years via Seminars, Professional & Technical Papers & Presentations.

125

SUCCESS

(For those people in the Quality, Reliability & Safety business)

The secret of success is to stay cool and calm on top and paddle like HELL underneath !

SUCCESS IS.....

Success means something different to each person.
To me it means self-esteem, which is the best defense against petty gossip, envy, unfair criticism
and naked lies.

Success means having intelligent friends & honest
critics. It means having the ability to laugh at
yourself and feel real pain when someone good is
in trouble.

Success is appreciating the best in others and ack-
nowledging the shortcomings in ourselves. And most
of all, Success is knowing that you have enriched
the life of even one person because something you
said or did elevated that person in some small way.
This, is not only Success, this is to have Succeeded!

Success can only be defined by you for you, since
you are a one-of-a-kind and unique in every respect.
Success is what you want it to be. Go for it!

Success is 'Motivated People' working in concert as
a team of individual contributors putting out 125%
for their company. Success is spelled...Y...O...U!

Success is....
Motivated People By Dr. Carter

A NEW DAY.......

This is the beginning of a new day. God has
given me this day to use as I will. I can
waste it....or use it for good, but what I
do today is important, because I am exchang-
ing a day of my life for it! When tomorrow
comes, this day will be gone forever, leav-
ing in its place something that I have traded
for it. I want it to be gain, and not loss;
good, and not evil; success, and not failure;
in order that I shall not regret the price
that I have paid for it.

                         By Dr. Heartsill Wilson

From: Southwestern Bell

Comment by Dr. Carter: Truer words will be hard
to find. Consider these very carefully and guide
yourself accordingly on your planned path of personal growth, development, achievement & success.

                         Dr. Chuck Carter

Quality

- Means Different Things to Different People.
- Relates the Features and Characteristics of a Product or Service to the Ability of that Product or Service to Satisfy Stated or Implied Needs.
- Is often referred to as a Degree of Excellence: Good, Perfect, Marginal, Outstanding, Poor, Bad, Sub-Standard, etc.
- Refers to Satisfying 'Needs,' and for those needs, Doing Things Right the First Time, without Replacement, Rework or Repair.
- Is Cost Effective. There is no Lower Cost than Doing the Job Perfectly One Time.
- Means Customer Satisfaction: On Time, at the Lowest Total Cost, per the Contract or Off-The-Shelf.
- Must be Controlled: Hence Q.C.
- Must be Assured: Hence Q.A.
- Assurance: To give Confidence, to Remove Doubt, Uncertainty or Worry.
- To Ensure: To make Sure, Certain or Safe.
- To Insure: To make Arrangements for Indemnification.

- Control: An Evaluation to Check, Test, or Verify in order to indicate where a Corrective Response is called for.
- Of Management (Leadership): The Act of Guiding, Directing or Managing.
- Control (Stability Sense): A State of a Process in which the Variability is Attributable to a Constant System of Chance Causes.
- Management: The Key Role of Top Management is Providing the Direction, Support, Policy, and Commitment to Quality.
- System: Describes Organization Structures, Responsibilities, Procedures, Processes and Human Resources for Implementing the Quality Management System of Operation.
- Plan: The Document Setting Out The Specific Quality Practices, Resources, and Activities Relevant to a Particular Product, Process, Service, Contract or Project.
- Total Quality Assurance: All Those Planned or Systematic Actions Necessary to Provide Adequate Confidence that a Product or Service will Satisfy the Given Requirement for Quality.
- Total Quality Control: All The Operational Techniques and The Activities Used to Fulfill Requirements of Quality.
- Total Quality Management: Every Department in the Company, All of Its Suppliers and Vendors, All of Its Management Team from Chairman of the Board on Down, and All of the Line, Staff and Total Human Resource Population as "Partners" Dedicated and Committed to Quality, Reliability, Safety and Integrity of Everything They Do Everyday in every Way, Quality Without Compromise.
- Quality: The Only Way to Productivity, Profitability and Success in the Market Places of the World.

References: ANSI/ASQC A3 - 1987
Dr. C. L. Carter, Jr., P.E.
Inspection Division Newsletter, Vol. 20-1989

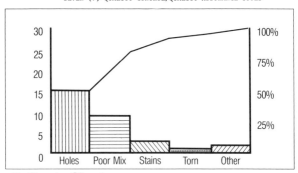

1 **Pareto Chart** — A sorted vertical bar graph which helps to determine which problems to solve in what order.

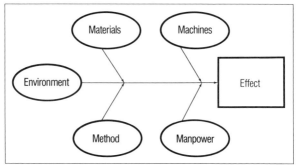

2 **Cause and Effect Diagram** — Represents the relationship between some 'effect' and its possible 'causes'.

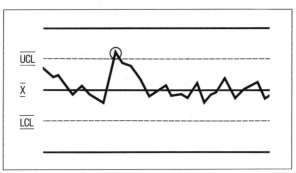

3 **Control Chart** — A time ordered chart with statistically determined upper and lower limits on either side of the process average.

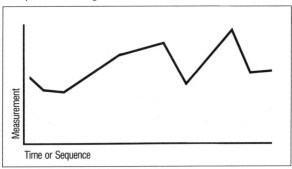

4 **Run Chart** — Display of observation points over a specified time period.

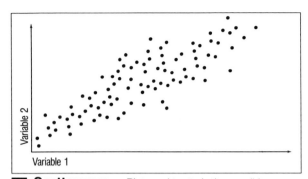

5 **Scattergram** — Plot used to study the possible relationship between one variable and another.

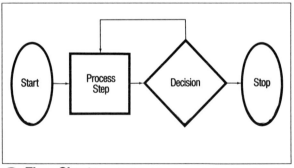

6 **Flow Charts** — A pictorial representation showing all the steps of a process.

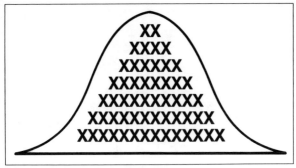

7 **Histogram** — Displays a distribution of data values.

Reference: Sanden International (USA), Inc.
World Class Mfg. Facility
Wylie, Texas

Comment By Dr. Carter: I had the pleasure of touring this Exceptionally
Fine Facility in October, 1990 as a part of THE
TEXAS QUALITY CONSORTIUM training meeting. It is
indeed WORLD CLASS QUALITY, INCLUDING Management,
The Total Quality Team, The Robots, The Processes,
The Facilities and The Environment. Very, very
Well Done!

SUPERCONDUCTING SUPERCOLLIDER

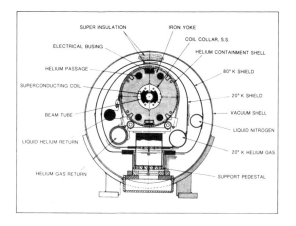

The \$8 Billion +++ Superconducting Supercollider Cross Section of the Atomic Particle Path.

Reference: Engineering Times
          National Society of Professional Engineers

INSPECTION TOOL AND EQUIPMENT READY-REFERENCE AND *Identification Chart* WITH SUGGESTED CALIBRATION INTERVALS AND PRACTICAL ACCURACY LIMITS.

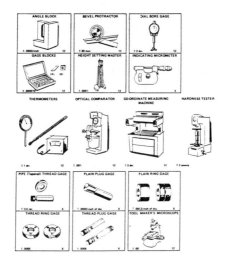

PRACTICAL ACCURACY LIMITS

The practical accuracy limits, as shown in the lower left hand corner of each illustration, are general and are wholly dependent on the specific instruments, the operator, and the conditions under which the tool or gage is used. The limits shown designate an area of possible disagreement when independent measurements are made by different qualified inspectors, using different but accurate tools under proper environmental conditions.

INSTRUCTIONS FOR USING THIS CHART

This chart has been prepared to assist Quality Assurance, Manufacturing, and Purchasing personnel in the identification of most commonly used inspection equipment. Generic names are used. You can write in proprietary nomenclatures if necessary. However, we recommend that the use of proprietary or trade names be discouraged.

The calibration intervals and accuracy limits are shown as guide lines. Should you desire to make changes, we suggest you draw a pencil line through the red numbers and write in your own numbers as dictated by conditions of use and past experience.

Reference: "Quality Assurance Workmanship Standards and Training Manual - By Dr. Carter"

CALIBRATION INTERVALS

The suggested initial Calibration Interval (in months) is shown in the lower right hand corner of each block. The intervals shown are those most commonly used. They should be modified to compensate for condition & frequency of use.

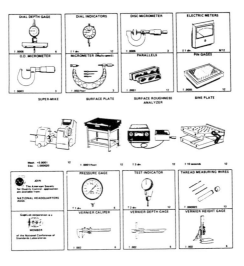

Courtesy of: GAGE LAB CORPORATION
Huntington Valley, Penna. 19006

Note: ISO-9000/Q-90 Standards Require Calibrated Tools, Gages & Test Equipment as Traceable to Nationally Recognized Standards, i.e. N.I.S.T., etc.

THE METRIC SYSTEM

The Metric System will be used in the U.S.A. in more and more
companies over the next few years. Many corporations, businesses,
and governmental agencies are switching to a dual or 100% metric
system. It will take many years for total conversion. During
this period of time, you will probably be working with the old
and new units of measure, weight and temperatures. Thus, we
all have the need to know and understand "The Metric System".
It will be a burden and a bother for awhile, but in time, you
will find it really isn't too difficult. For those in Engineering and related technical career fields, this material
may be very basic since you are probably working with "The
Metric System" in one way or another.

THE BASIC METRIC TERMS

<u>Celsius</u> (Centigrade) (Abbreviation = C)

You are probably already hearing the word Celsius (centigrade)
used on your daily television and radio weather reports, so lets
start with temperature measurement.

1. Water freezes at 0 degrees celsius   (32° Fahrenheit)

2. Water boils at 100 degrees celsius   (212° Fahrenheit)

3. Your temperature is "Normal" at 37° celsius   (98.6° Fahrenheit)

4. You would need to check with the doctor at
   38.33° celsius   (101° Fahrenheit)

5. A comfortable outdoor temperature will be
   21.11° celsius   (70° Fahrenheit)

Reference: "Quality Assurance Workmanship Standards
            and Training Manual - By Dr. Carter"

6. Air Conditioned Interiors will be
   comfortable at about 23.89° celsius   (75° Fahrenheit)

7. During the hot summer months you will
   really feel the heat at 37.78° celsius   (100° Fahrenheit)

8. During the cold winter months you will
   really feel the cold at -17.78° celsius   (0° Fahrenheit)

How to Convert Celsius Degrees to Fahrenheit

First, you must have a known celsius degree temperature

EXAMPLE: 20° celsius (which we can prove is 68° F)
          20 x 9 = 180 ÷ 5 = 36 + 32 = 68° F
                 20° C = 68° F

How to Convert Fahrenheit Degrees to Celsius

First, you must have a known Fahrenheit degree temperature

EXAMPLE: 86° Fahrenheit (which we can prove is 30° C)
          86 - 32 = 54 ÷ 9 = 6 x 5 = 30° C
                 86° F = 30° C

Chart of Additional Conversions

   Celsius to Fahrenheit

0	=	32
10	=	50
20	=	68
30	=	86
35	=	95
37	=	98.6
40	=	104
50	=	122
60	=	140
70	=	158
80	=	176
90	=	194
100	=	212

Fahrenheit	to	Celsius		Fahrenheit	to	Celsius
0	=	-17.78		95	=	35
5	=	-15		98.6	=	37
10	=	-12.22		100	=	37.78
20	=	- 6.67		101	=	38.33
32	=	0		102	=	38.89
60	=	15.56		103	=	39.44
65	=	18.33		104	=	40
70	=	21.11		105	=	40.55
75	=	23.89		110	=	43.33
80	=	26.67		150	=	65.55
85	=	29.44		200	=	93.33
90	=	32.22		212	=	100

Gram   (Abbreviation = G)

Your relationship to the Gram will be most familiar through
doctors, nurses, pharmacists, drug stores and bottled goods.
The Gram has been used for years to accurately measure medicine. One aspirin is about 5 Grams. A normal-size cig-
arette is about one Gram. You will see Ounces and Grams
stated on the same container. One Ounce is 28.35 Grams and
One Gram is .035 Ounces.

Chart of Additional Conversions:

Grams	=	Ounces		Grams	=	Ounces
1	=	.035		15	=	.529
2	=	.071		20	=	.705
3	=	.106		25	=	.882
4	=	.141		50	=	1.764
5	=	.176		75	=	2.646
6	=	.212		100	=	3.527
7	=	.247				
8	=	.282				
9	=	.317				
10	=	.353				

```
Ounces  =  Grams

  .1   =     2.835
  .2   =     5.67
  .3   =     8.505
  .4   =    11.34
  .5   =    14.175
  1.   =    28.35
  2.   =    56.7
  3.   =    85.05
  4.   =   113.4
  5.   =   141.75
  6.   =   170.1
  7.   =   198.45
  8.   -   226.8
  9.   =   255.15
 10.   =   283.5
```

Milligram (Abbreviation = mg)

A Milligram is very light in weight and you really would have
difficulty in feeling anything that light. Measuring in
Milligrams is highly accurate. Normally, you will not come
in contact with this unit of measure, unless.... work is
highly technical and highly accurate measurements are required.

Chart of Conversions:

Milligrams	=	Ounces		Ounces	=	Milligrams
1	=	.000035		.01	=	283.495
5	=	.000176		.05	=	1417.476
10	=	.000353		.10	=	2834.952
15	=	.000529		.15	=	4252.428
20	=	.000705		.20	=	5669.904
40	=	.001411		.40	=	11339.807
60	=	.002116		.60	=	17009.707
80	=	.002822		.80	=	22679.607
100	=	.003527		1.	=	28349.523
				10.	=	283495.21

<u>Kilogram</u> (Abbreviation = kg)

You will relate to the Kilogram since it equates to pounds and you are familiar with how much you weigh in pounds. One kilogram is a little over two pounds. If your wife weighs 100 pounds, she will appreciate the fact that she only weighs 45.36 Kilograms. If you weigh 150 pounds, you will tip the scale at 68.04 Kilograms.

<u>Chart of Additional Conversions</u>:

Kilograms	=	Pounds		Pounds	=	Kilograms
1	=	2.205		1	=	.454
5	=	11.023		5	=	2.268
10	=	22.046		10	=	4.536
15	=	33.069		15	=	6.804
20	=	44.092		20	=	9.072
50	=	110.231		50	=	22.68
100	=	220.462		100	=	45.36
200	=	440.924		150	=	68.04
1000	=	2204.623		200	=	90.72
				250	=	113.4
				1000	=	453.6

<u>Litres</u> (Abbreviation = l)

You can relate to Litres since it equates to quarts and gallons which will be well known to you for items such as milk, oil and gas, ice cream, etc. One Litre of oil is slightly more than one quart of oil.

Chart of Conversions:

Litres	=	Quarts
1.	=	1.0567
2.	=	2.1134
5.	=	5.2834
8.	=	8.4535
10.	=	10.5669

Litres	=	Gallons
1.	=	.264172
2.	=	.52834
5.	=	1.32086
8.	=	2.11338
10.	=	2.64172

Quarts	=	Litres
1.	=	.9464
2.	=	1.8927
5.	=	4.7318
8.	=	7.5708
10.	=	9.4635

Gallons	=	Litres
1.	=	3.7854
2.	=	7.5708
5.	=	18.927
8.	=	30.2832
10.	=	37.854

Millilitres (Abbreviation = mm)

It takes 1000 millilitres to make a litre. Thus, you can see that millilitres are very small. Millilitres equate to Liquid Ounces and are mostly used to measure medicine and other items where accuracy and precision are important. It takes a little more than 946 millilitres to make one quart.

Chart of Conversions:

Millilitres	=	Liquid Ounces
1	=	.034
2	=	.068
5	=	.169
10	=	.338
20	=	.68
100	=	3.38
1000(litre)	=	33.81

Liquid Ounces	=	Millilitres
1	=	29.574
2	=	59.147
5	=	147.868
10	=	295.735
20	=	591.471
25	=	739.338
32(quart)	=	946.353

<u>Metre</u> (Abbreviation = m)

You are familiar with yards and feet as a unit of measure. The metre equates to yards and feet. Actually, one metre is just a little over one yard. If you have been to a professional base-ball game this year, you will recognize the metre as the dual measurement to the outfield walls.

<u>Chart of Conversion:</u>

Metres	=	Feet		Feet	=	Metres
1	=	3.281		1	=	.305
2	=	6.562		2	=	.61
5	=	16.404		5	=	1.524
8	=	26.247		8	=	2.438
10	=	32.808		10	=	3.048

Metres	=	Yards		Yards	=	Metres
1	=	1.094		1	=	.914
2	=	2.187		2	=	1.829
5	=	5.468		5	=	4.572
8	=	8.749		8	=	7.315
10	=	10.936		10	=	9.144

Centimetre (Abbreviation = cm)

The Centimetre equates to inches and this should be somewhat familiar if you have seen the new dual reading rulers. One inch equals about 2½ centimetres. There are 100 centimetres in a metre.

Chart of Conversions:

Centimetres	=	Inches
1	=	.394
3	=	1.181
5	=	1.969
10	=	3.937
15	=	5.906
30	=	11.811
75	=	29.528
100	=	39.37

Inches	=	Centimetres
1	=	2.54
3	=	7.62
5	=	12.7
10	=	25.4
15	=	38.1
30	=	76.2
75	=	190.5
100	=	254.

<u>Millimetre</u> (Abbreviation = mm)

Millimetres are small and are also found on the dual reading
scales and rulers. There are a little more than 25 millimetres
in one inch. The familiar paperclip is about 8 millimetres wide.

<u>Chart of Conversions</u>:

Inches = Millimetres

1	=	25.4
2	=	50.8
5	=	127.0
8	=	203.2
10	=	254.0

Decimal Inches	=	Millimetres	Millimetres	=	Decimal Inches
.001	=	.0254	1	=	.03937
.002	=	.0508	2	=	.07874
.005	=	.127	5	=	.19685
.008	=	.2032	7	=	.27559
.01	=	.254	10	=	.3937
.03	=	.762	50	=	1.9685
.06	=	1.524	100	=	3.937
.09	=	2.286	250	=	9.8425
.1	=	2.54	500	=	19.685
.2	=	5.08			
.5	=	12.7			
.8	=	20.32			
1. inch	=	25.4			

Kilometre (Abbreviation = km)

The Kilometre equates to miles. If you have been to Mexico or other foreign countries, the highway signs will give the distance to the next town or city in kilometres. Perhaps your state now has dual highway signs, giving you miles and kilometres. 8 kilometers is almost five miles.

Chart of Conversions:

Kilometres	=	Miles		Miles	=	Kilometres
1	=	.621		1	=	1.609
2	=	1.243		2	=	3.219
3	=	1.864		3	=	4.828
4	=	2.485		4	=	6.437
5	=	3.107		5	=	8.047
10	=	6.214		10	=	16.093
15	=	9.321		15	=	24.14
30	=	18.641		30	=	48.28
60	=	37.282		60	=	96.56
90	=	55.923		90	=	144.84
500	=	310.686		500	=	804.7

THE METRIC SYSTEM BASIC TERMS

Metre.....(Distance)

Litre.....(Liquid)

Gram......(Weight)

Celsius...(Temperature)

Kilo = 1 thousand	A Kilogram is 1000 grams
Centi = 1 hundredth	A Centimetre is .01 of a metre
Milli = 1 thousandth	A Millilitre is .001 of a litre

QUALITY CONSORTIUMS

Training & Development Consortiums are getting to be popular
in several states and I hope this concept continues to grow
in the 1990's and beyond. Formed to help companies become
"QUALITY COMPANIES" through group Training & Development Programs; Self-teaching/Self-help and Sharing one with the other,
over extended periods of time i.e. 2,3,4,5 years of planned,
systematic training, learning, performing, progressing and
achieving is working for those who are involved and plan to
succeed via **"The Quality Route to Success"**.

THE TEXAS QUALITY CONSORTIUM is one that I am familiar with
as an involved Trainer and as a representative of one of the
member companies. Mr. Warren Hogan formed Hogan & Associates,
Inc. in 1987 as "A Training & Consulting Company in Total Quality". He had been in Executive Management positions with Texas
Instruments and with Airborn Electronics as President & CEO,
always as a Prime Mover for Quality. The Texas Quality Consortium was formed with several small companies primarily in the
Electronic Fields who wanted to Learn, Grow & Develop their
knowledge of Quality to the point of Implementing Total Quality Management Systems in their companies. They collectively
agreed with the name and Hogan & Associates, Inc. would lead
and provide the Planned, Systematic Training to allow all of
the members to achieve their goals at reasonable low costs to
be paid as you go. See Prof. Reference Library for Contact Info.

To make a successful story shorter....There are now four(4)
groups of 8 to 12 companies and at this point there are 38
firms. Some have dropped-out, some have come back, some are
new. To become a Member, each must be **SERIOUS ABOUT QUALITY.**
As a GROUP, the companies undergo Training, Development & Progress Sharing over a period of two(2)years. This provides companies with mutual reinforcement, individual & collective assistance and helps reduce costs. The consortium provides monthly
Executive-Level Breakfast Meetings; monthly formal training
sessions held at the University of Texas at Dallas Campus; individual company counseling & ongoing reinforcement; assistance
with recruiting, procurement, and more. Yes, this is a success
story. But there is more....Two of the companies are planning
to **GO FOR THE BALDRIGE AWARD! That is really Fantastic!**

There are More Consortiums..... Each formed for about the same
reasons: **"QUALITY IS THE ROUTE TO SUCCESS".** They are as follows:

QUALITY NEW JERSEY was formed in 1989 and now consists of 20
participating organizations. Contact: Stanley Marash(201)548-0600

THE CENTER FOR QUALITY MANAGEMENT is in Massachusetts and consists of approximately 7 companies. Contact: Ray Stata, President
of Analog Devices, Inc.

THE OHIO PARTNERSHIP consists of The Ohio Manufacturer's Assoc.
and the Ohio Dept. of Development. Contact: Eric Burkland at
(614) 224-5111.

147

Dr. J. Juran's 10 Steps To Quality Improvement

1. Build Awareness of the need & opportunity for improvement.

2. Set Goals for Improvement.

3. Organize to Reach the Goals.(Establish a Quality Council, Identify Problems, Select Projects, Appoint Teams, Designate Facilitators)

4. Provide Training.

5. Carry out Projects to resolve problems.

6. Report Progress.

7. Give Recognition.

8. Communicate Results.

9. Keep Score.

10. Maintain momentum by making annual improvement a part of the regular systems & processes of the company.

Dr. W. E. Deming's 14 Steps To Survival

1. Create Constancy of Purpose Toward Improvement.

2. Adapt New Philosophy: Take on Leadership for Change.

3. Eliminate Inspection: Build Quality.

4. Award Business Based on Loyalty & Trust.

5. Improve Constantly.

6. Institute Training on the Job.

7. Institute Leadership - Leaders Help People.

8. Drive Out Fear.

9. Break Down Barriers Prohibiting Team Development.

10. Eliminate Slogans.

11. Eliminate Standards, MBO, etc. - Use Leadership.

12. Remove Barriers Which Rob Employees of Pride & Satisfaction.

13. Begin Vigorous Program of Education/Self-Improvement.

14. Transformation Is Everybody's Job.

Phil Crosby's 14 Step Program

1. Management Commitment to Change and to Quality.

2. Form Quality Improvement Teams in each department.

3. Determine & Measure Quality in each department.

4. Evaluate & Report Cost of Quality. Estimate; then report facts.

5. Quality Awareness must be made & shared with all employees.

6. Take Corrective Action after you seek, search & find problems.

7. Develop Zero Defects Program with Q.I.Teams leading/implementing.

8. Supervisor/Management Training & Development is a Must.

9. Zero Defects Day must be a major event to display New Attitudes.

10. Attainable, Measurable Goal Setting by all employees.

11. Error Cause Removal by all employees, management & staff.

12. Highly Visible Recognition Program with Rewards for contributors.

13. Quality Councils consisting of Quality Professionals & Q.I.Team Chairpersons should be formed for Communication & Program upgrading.

14. Repeat & Do It All Over Again after first 13 are accomplished.

TOTAL QUALITY CONTROL DEFINED
By Dr. A.V. Feigenbaum

"An effective system for integrating the quality-development, quality-maintenance, and quality improvement efforts of the various groups in an organization so as to enable marketing, engineering, production, and service at the most economical levels which allow for full customer satisfaction."

Dr. C. L. Carter's 5 Phase Quality Program

1. Prevention: The Goal is to have a Formal Program to Prevent Quality & Related Problems.
2. Control: Process Controls; Control to Workmanship Standards.
3. Assurance: Check & Assure your Products, Processes, Procedures.
4. Assistance: Assist, Train & Develop Your Employees: Major Assets.
5. Auditing: Conduct unbiased Audits: Anything, Anywhere, Anytime.

Note: See Page 91,92,93 for Specifics of Dr. Carter's Program.

Malcolm Baldrige
**National
Quality
Award**

Managed by:
United States Department of Commerce
National Institute of Standards and Technology
Gaithersburg, MD 20899
Telephone (301) 975-2036
Telefax (301) 948-3716

Administered by:
The Malcolm Baldrige National Quality
Award Consortium, Inc.

P.O. Box 443
Milwaukee, WI 53201-0443
Telephone (414) 272-8575
Telefax (414) 272-1734

P.O. Box 56606, Dept. 698
Houston, TX 77256-6606
Telephone (713) 681-4020
Telefax (713) 681-8578

*"The improvement
of quality in
products and the
improvement of
quality in service –
these are national
priorities as
never before."*

George Bush ,

*"The success of the
Malcolm Baldrige
National Quality Award
has demonstrated
that government and
industry, working together,
can foster excellence."*

**Robert Mosbacher
Secretary of Commerce**

MALCOLM BALDRIGE
NATIONAL QUALITY AWARD
FACT SHEET

MALCOLM BALDRIGE NATIONAL QUALITY AWARD

Public Law 100-107, the Malcolm Baldrige National Quality Improvement Act of 1987, signed by President Reagan on August 20, 1987, established an annual U.S. National Quality Award. The purposes of the Award are to promote quality awareness, to recognize quality achievements of U.S. companies, and to publicize successful quality strategies. The Secretary of Commerce and the National Institute of Standards and Technology (NIST, formerly the National Bureau of Standards) are given responsibilities to develop and administer the Awards with cooperation and financial support from the private sector.

THE AWARDS

Up to two Awards may be given each year in each of three categories:

- ▸ manufacturing companies or subsidiaries
- ▸ service companies or subsidiaries
- ▸ small businesses

NOTE: Fewer than two Awards may be given in a Category if the High Standards of the Award Program are not met. The First Awards were presented in 1988 by President Reagan.

1988 Award Winners:

Motorola, Inc. The Entire Corporation. Manufacturing Company Category.
Products: Electronics/Communications & Related.

Westinghouse Electric Corporation's Commercial Nuclear Fuel Division: Manufacturing Company Category. Product: Nuclear Fuel.

Globe Metallurgical, Inc. The Entire Corporation. Small Business.
Products: Ferroalloys & Silicon Metals

1989 Award Winners: Presented by President George Bush 11/2/89

Milliken & Company: The Entire Company. Manufacturing Company Category.
Products: Milikens 28 businesses produce more than 40,000 different textiles & chemical products.

Xerox Business Products & Systems: The Entire Company. Manufacturing Company Category.
Products: Over 250 types of Document Processing Equip.

Note: In 1988 there was No Service Company Winner.

In 1989 there was No Service Company Winner and No Small Business Winner.

151

1990 Award Winners: Announced in October by Commerce Secretary
                    Robert A. Mosbacher to be Presented by President Bush.

   Cadillac Motor Car Division of General Motors. Manufacturing Category.
                    Product: Automobiles

   IBM Rochester, Minn. Division. Manufacturing Category.
                    Product: Minicomputers

   Federal Express Corp. The Entire Corporation. Service Category.
                    Service: Air Express Delivery. Over 1.3 million ship-
                    ments daily worldwide.

   Wallace Co. Inc., Houston, Texas. The Entire Company of 280 people.
                    Small Business Category as Distributor.
                    Products: Pipe, Valves, Fittings for Petrochemical &
                               Refining industry along Gulf Coast of Texas.

Note: In 1990 the First Award for 'Services' was made.
      In 1990, 160,000 requests for applications were made and 97 APPLIED.
      Of those, 45 were in Mfg.Cat. where 6 on-site visits were
      conducted. There were 18 Service Cat. applicants with 3 on
      site visits made. Small Business had 34 applicants with 3
      visits made.

Note: PLEASE KEEP TRACK AND RECORD THE WINNERS FROM YEAR TO YEAR....
                         (Maximum 2 in Each Cat. each Year)

1991 Award Winners:

   Mfg. Cat.
   Mfg. Cat.
   Sm. Bus. Cat.
   Sm. Bus. Cat.
   Service Cat
   Service Cat.

1992 Award Winners:

   Mfg. Cat.
   Mfg. Cat.
   Sm. Bus. Cat.
   Sm. Bus. Cat.
   Service Cat.
   Service Cat.

1993 Award Winners;

   Mfg. Cat.
   Mfg. Cat.
   Sm. Bus. Cat.
   Sm. Bus. Cat.
   Service Cat.
   Service Cat.

The Man In The Glass

When you get what you want in your struggle for self
 And the world makes you king for a day
Just go to the mirror and look at yourself
 And see what that man has to say.

For it isn't your father or mother or wife,
 Whose judgement upon you must pass.
The fellow whose verdict counts most in your life
 Is the one staring back from the glass.

Some people might think you're a straight-shooting' chum,
 And call you a wonderful guy.
But the man in the glass says you're only a bum
 If you can't look him straight in the eye.

He's the fellow to please, never mind the rest
 For he's with you clear to the end.
And you've passed your most dangerous test
 If the guy in the glass is your friend.

You may fool the whole world down the pathway of years
 And get pats on the back as you pass.
But your final reward will be heartache and tears
 If you've cheated the man in the glass!!!

— *Author Unknown*

Comment by Dr. Carter: Not wanting to change the wordage
of the above, please personalize this by using the word
'Person or Woman' as may be the case.

Also, as we all look at all the Major Problems we now
have in the United States regarding Fraud, the lack of
Personal Integrity, Cheating, Deception, Greed, Untruths,
Disunity of the Family, Child Abuse, Murder, etc., all of
which has touched all levels of Government, Business, Industry, Banking, Saving & Loans, Church/Religion, Police,
Judge & Jury to name a few, I would refer each and every
one to the last two lines..."But your final reward will be
heartache and tears, If you've cheated the man in the glass!!!"

Restrictions on Eligibility

The intent of Public Law 100-107 is to create an Award process incorporating rigorous and objective evaluation of the applicants' total quality systems and underlying products and services. Award recipients are to s[...] as appropriate models of total quality achievement for other United State[...] companies. Customer satisfaction is to play a major role in the examinatio[...] Site visits are required to verify descriptions given in written applications.

The nature of some companies' activities are such that the central purpos[...] and requirements of Public Law 100-107 cannot be fulfilled through their participation in the Award program; companies or subsidiaries whose businesses cannot fulfill these purposes are not eligible. Specifically, three restrictions apply:

1. A company or its subsidiary is eligible only if the quality practices associated with all major business functions of the applicant are inspec[...] in the U.S. or its territories: One or more of the following three condition[...] must apply:
   - more than 50% of the applicant's employees must be located in the [...] or its territories, or
   - more than 50% of the applicant's physical assets must be located in [...] U.S. or its territories, or
   - more than 50% of the total quality management operations which underlie the products and services it delivers are conducted inside th[...] U.S. or its territories

   Note: The functions/activities of foreign sites must be included in the application report in the appropriate Examination Item.

2. For a subsidiary to be eligible, at least 50% of the customer base (dolla[...] volume for products and services) must be free of direct financial and li[...] organization control by the parent company. For example, a subsidiary [...] not eligible if its parent company or other subsidiary of the parent compa[...] are the customers for more than one-half of its total products and service[...]

3. Individual units or partial aggregations of units of "chain" organizations (such as hotels, retail stores, banks, or restaurants), where each unit performs a similar function, or manufactures a similar product, are not eligible.

Multiple-Application Restrictions

1. A subsidiary and its parent company may not both apply for Awards in th[...] same year.

2. Only one subsidiary of a company may apply for an Award in the same ye[...] in the same Award category.

Future Eligibility Restrictions on Award Recipients

1. If a company receives an Award, the company and all its subsidiaries are ineligible to apply for another Award for a period of five years.

2. If a subsidiary receives an Award, it is ineligible to apply for another Awar[...] for a period of five years.

3. If a subsidiary constituting of more than one-half of the total sales of a company receives an Award, neither that company nor any of its other subsidiaries is eligible to apply for another Award for a period of five years[...]

onfidentiality: All Applications are CONFIDENTIAL.

Applications or More Information, Please Write or Call:

National Institute of Standards & Technology, Room A[...]
Gaithersburg, Maryland 20899 Phone(301) 975-2036

155